Einstein, the Aether & Variable Rest Mass:

*Correcting misunderstandings that
have made relativity seem counterintuitive*

Jack Heighway

ISBN 978-1-61658-620-1

For all those whose work has been suppressed
by editors who deem it their duty
to protect the mainstream
from the contamination of new ideas

Contents

Special preface to the second edition intended to persuade physicists and others familiar with Einstein's theory of gravity that the correct interpretation of the theory must admit variable rest masses.

It has been my experience that physicists are strongly disinclined to accept the idea of variable rest masses. Unfortunately, in the first edition of this book, the topic is not addressed until page 50. I hope that this brief presentation will serve to persuade experts that the variable rest mass concept is sound and important. Others, I hope, taking for granted the mathematics, will be able to get the gist of the argument. The proper metric for the Schwarzschild field may be written

$$ds^2 = f^2 c^2 dt^2 - f^{-2} dr^2 - r^2 (d\theta^2 + \sin^2 \theta \, d\phi^2) \qquad f = \sqrt{1 - r_S / r}$$

$$r_S = 2GM / c^2$$

Consider a laser beam directed vertically from r_1 to r_2, ($r_1 < r_2$). The laser is presumed to operate continuously, and since the field is time-independent, the number of wavecrests in the interval between r_1 and r_2 will be constant. According to the metric, an observer at r_2 will record a decrease (red shift) of the laser frequency by the factor, $f(r_1)/f(r_2)$.

First suppose that the emission frequency of the source, and the frequency response of the receiver, are unaffected by the gravitational field. This cannot be the case, for it would imply that at r_2, wavecrests are being counted at a lesser rate than they are being emitted at r_1, which would mean that the number of wavecrests in the interval between r_1 and r_2 would increase with time, contrary to assumption.[1] So the characteristic frequencies of the instruments *must* depend upon their location in the field. Quoting Einstein,[2]

> "..an atom absorbs or emits light of a frequency which is dependent on the potential of the gravitational field in which it is situated."

Unstated by Einstein is a vital corollary:

> "In a time-independent field, light propagates with constant frequency, even though local observers, with their position dependent 'standards,' will not agree."

Obviously this is required in order that energy be conserved, as it must be in the time-independent Schwarzschild field. To this point, we may conclude that a new

---

[1] This argument was given by Einstein in his article of 1911: *Annalen der physik* **35,** translated as 'On the influence of gravitation on the propagation of light' in *The Principle of Relativity* 1952 (Dover) pp.97-108

[2] A Einstein, *Relativity*, (Crown Publishers, New York, 1952), 15th ed., Appendix III(c), pp.130-131

scheme for time keeping is to be desired, a scheme in which the frequency of a freely moving light wave will be constant, independent of the potential of the gravitational field. More of this later.

Next consider the annihilation of a ground-state para-positronium system occurring at radial coordinate, r, deep in the Schwarzschild field. Distant observers, whose frequency standards are not significantly influenced by the field, will measure the frequency of the annihilation gamma rays to be red shifted by the factor, $f(r)$. According to the above corollary, the total energy of the gamma rays is unchanged in flight, and is exactly equal to the rest mass energy of the electron-positron pair. The inescapable conclusion is that the rest mass of the pair was reduced by the factor $f(r)$. So rest masses *are* reduced in a gravitational field.

Now consider the consequences of such rest mass reduction. According to the *Strong Equivalence Principle* (SEP),[3] all standards of time intervals and length intervals defined in terms of physical systems must vary together (if they do in fact vary at all). Niels Bohr's 1913 quantum description of the hydrogen atom provides ideal standards for characteristic frequencies (periods) and dimensions of physical objects:

The Rydberg period: $P_R = h^3 / m_e e^4$, and the Bohr radius: $a_0 = h^2 / m_e e^2$

A reduction in rest mass by the factor f will result in an increase in the Rydberg period by the factor f^{-1}. Thus it may be concluded that rest mass reduction is the cause of the red shift.

But the reduction of rest mass also has the effect of increasing the Bohr radius by the factor f^{-1}. This means, in accord with the SEP, that the length of all physical objects, including measuring rods, will increase by f^{-1}. It follows that the presently accepted proper geometry (based as it is, in effect, on measurements made with such rods) is incorrect, from which it may be concluded that a new scheme for space measurements, as well as for time keeping, is desirable.

Because of the slowing of clocks and the elongation of measuring rods, measured time and distance intervals will both be underestimated by the factor f. The

---

[3] "In a freely falling, non-rotating laboratory, the local laws of physics take on some standard form, including a standard numerical content, independent of the position of the laboratory in space and time." R H Dicke *The Theoretical Significance of Experimental Relativity* (Gordon and Breach, New York, (1965) p.4

It is important to note that the SEP does not assert that nothing can change: it does allow that the characteristics of instruments used for the measurement of time, length and mass (and perhaps fundamental 'constants' as well) may depend upon position in a gravitational field, provided that they vary together in such a manner as to satisfy the SEP.

correction to the metric is effected by simply multiplying ds^2 by the factor f^{-2}. Thus we introduce a new metric, distinguished by appending an asterisk, *.

$$ds*^2 = f^{-2}ds^2 = c^2dt^2 - f^{-4}dr^2 - f^{-2}r^2(d\theta^2 + \sin^2\theta\, d\phi^2)$$

This metric correctly describes the time and distance intervals in the Schwarzschild field. Importantly, it must *not* be thought of as a replacement for the proper metric. The proper metric correctly describes the kinematic behavior of particles with or without rest mass, whereas the new metric describes only the behavior of light and other zero-mass quanta.

The analysis presented above is well out of the mainstream. In order to understand how the present misinterpretation of the theory came about, it will be useful to examine the history of the development of Einstein's theory of gravity.

In his wonderful paper of 1911, Einstein invoked his *Equivalence Principle* [4] to predict two phenomena – the curvature of light rays, and the frequency reduction (red-shift) of atomic spectra in a gravitational field.

At that time, Einstein had no premonition that his eventual mathematical theory of 1915 would involve non-Euclidean geometry. Thus, in trying to understand the curvature of light rays, he quite naturally supposed that the gravitational field acted as if it possessed an index of refraction, so that the speed of light would be reduced in a gravitational field. (In terms of the Schwarzschild field $\hat{c} = f c$, the circumflex indicating what Einstein conceived to be the true value.)

Regarding the second prediction, which may be called the slowing of clock rates, Einstein immediately realized that there could be a conflict with the key postulate of special relativity, namely, that local observers will always measure the speed of light to be the same constant value. It was natural that he should immediately invoke his supposition regarding the reduced speed of light in order to 'rescue' special relativity.

But clearly, the slowing of clocks could be compensated for by an elongation of measuring rods. The time of flight would be underestimated, but so also would be the distance traveled, leaving the measured speed of light unaffected.

But again, in 1911, neither Einstein nor anyone else possessed an overarching theory describing the behavior of clocks and measuring rods.

This is how it came about that Einstein's *ad hoc* assumption that the speed of light is reduced in a gravitational field has become a part of the dogma of physics.

Note that one may write

---

[4] Uniform acceleration in flat space is equivalent to being stationary in a gravitational field.

$$ds^2 = \hat{c}^2 dt^2 - f^{-2}dr^2 - r^2(d\theta^2 + \sin^2\theta\, d\phi^2)$$

According to Einstein, and modern interpretation, half the bending of light rays is attributable to a non-Euclidean factor, f^{-2}, in the second term, and half is attributable to the reduced speed of light, $(\hat{c}/c)^2$. (Recall that Einstein's original prediction was just half the correct value, since at that time, he had not imagined that gravity would induce a non-Euclidean geometry.) In contrast, the new metric ascribes the entire bending to a 'more curved' non-Euclidean geometry, principally, f^{-4}, in the second term:

$$ds^{*2} = c^2 dt^2 - f^{-4}dr^2 - f^{-2}r^2(d\theta^2 + \sin^2\theta\, d\phi^2)$$

It is remarkable that Einstein did not attempt to explain the slowing of the speed of light, or the frequency reduction of atomic spectra in a gravitational field. Incredibly, many contemporary physicists presume to explain both the slowing of clocks and the reduction of the speed of light with a single, stunningly mystical idea – asserting that *"the flow of time itself slows"* in a gravitational field.

The origin of this singularly unscientific concept can be traced back to Einstein's 1905 re-interpretation of the relativity theory of Lorentz, Larmor and Poincaré, in which he relegated the aether concept to the trash bin of superfluous concepts, and thus introduced the idea of the relativity of time. Two observers passing one another each claim the other's clock runs slowly: without the aether, [5] the symmetry is unbroken, and it hence it seems to follow that time is relative – the other guy's clock doesn't run slowly for any discernable cause – it's just that time is relative!

Crucially, this 'explanation' for the slowing of clock rates and the reduction of the speed of light proves to be incorrect, since it implies that the presently accepted proper spatial metric correctly describes the geometry of space in and around the region of photon orbits in the Schwarzschild field. But at the photon orbit radius, centrifugal force vanishes and inside this region, the *direction* of centrifugal force is *reversed!* (The Abramowicz effect.[6]) This behavior is incomprehensible in terms of proper geometry, but is an obvious consequence of the new geometry, as we now prove.

In the new metric, the area of a centered sphere is equal to

---

[5] The existence of the aether can be proved – at least in a mathematical sense, by a consideration of an imaginary – but possible – universe, the topology of which is that of a flat three-torus.(cf, pp. 13-14)

[6] Abramowicz M A and Lasota J-P (1974) "A note of a paradoxical property of the Schwarzschild solution" *Acta Physica Polonica* B5, 327

Abramowicz M A and Lasota J-P (1986) "On traveling round without feeling it and uncurving curves" *Am. J. Phys.* **54** (10) 936-38

Abramowicz M A (1992) "Relativity of inwards and outwards: an example" *Mon. Not. Astr. Soc.* 710-18,

$$4\pi r^2 f^{-2} = 4\pi r^2 (1 - r_S / r)^{-1},$$

Differentiating with respect to r, one has

$$d / dr [r^2 (1 - r_S / r)^{-1}] = (2r - 3r_S)(1 - r_S / r)^{-2}$$

Thus the area of the sphere is not a monotone function of r : it has a minimum at $r = 3/2\ r_S$, the locus of the photon orbits. Furthermore, for $r < 3/2\ r_S$, the area of the sphere increases, so that the surface of a centered sphere, viewed from the 'outside,' will be concave rather than convex. Thus, regarding the Abramowicz effect, the correct geometry shows that for $r < 3/2\ r_S$, the direction of centrifugal force obeys the usual pattern: the force is directed from the concave side to the convex side of the circle on which a body is constrained to move.

I believe that the most important consequences of the variable rest mass interpretation follow from the proof that light moves with constant frequency along the spatial geodesics of the true geometry. Light does not respond directly to gravity, although, understandably, it must be influenced by the non-Euclidean geometry induced by the gravitational field. Conversely, the energy of freely moving light does not act as a source of gravity. (Wheeler's Geon[7] is an impossibility.)

Thus, mass and energy are not in every way equivalent. *Gravity acts only on objects with rest mass!* [8] The implications are deep and far-reaching. In particular, the massless gravitons quantizing the gravitational field would not interact with that field, nor with one another. This implies that a quantum theory of gravity need not be intractably non-linear.

Of lesser importance are conclusions regarding black holes. A collapsing star does not suffer an infinite compression, producing a singularity hidden behind an event horizon; when its proper circumference becomes less than $(9\pi / 4)r_S$, it produces a sort of rupture in space, forming a connection to a distinct infinite realm, *innerspace*, into which it falls. This space is connected to our familiar universe by a flat sphere, the *stenosphere*, the locus of the photon orbits. The event horizon is the infinity of innerspace, and thus nothing has ever fallen, or ever will fall, through the event horizon. There is no worrisome loss of information, and the terrifying singularity at $r = 0$ is a non-problem.

---

[7] John Archibald Wheeler, Phys. Rev. 97, 511 - 536 (1955)

[8] Virtual photons *do* contribute to the rest mass of charged particles, but these photons are *confined*, and such confined photons do possess an effective mass, unlike *free* photons. (cf, Appendix V, p.100)

This second edition also incorporates an analysis of the manner in which rest masses evolve with time, thus completing the treatment of the cosmology according to the Variable Rest Mass interpretation. The analysis is based upon the conservation of momentum, which is implied by the assumption of the large scale homogeneity of the universe. Einstein's equations are developed and solved, showing that rest masses increase as a simple exponential function of world time.

Preface to First Edition

In writing this book I have had two purposes in mind. The first is to remove most of the mystery which has long permeated the subject, by providing a basis upon which the intelligent reader may achieve a real understanding. The second is to prove that the correct interpretation of Einstein's theory of gravitation must comprehend variable rest masses.

Part I deals with special relativity, showing that its non-intuitive aspects stem from a failure to recognize the implications of the prescribed method for synchronizing clocks. The existence of the aether is proved, at least in a mathematical sense, by a consideration of an imaginary – but possible – universe, the topology of which is that of a flat three-torus. Several seemingly mysterious phenomena of special relativity are analyzed in detail from the aetherist standpoint and are thus made comprehensible.

Part II presents the correct interpretation of Einstein's theory of gravity, in which variable rest masses plays a dominant role. The mathematics of Einstein's theory of gravity is accepted, but the conventional interpretation of the theory is shown to be in error. This conclusion follows from a careful investigation of the cause of gravitational slowing of clocks, which reveals that the reduction of rest mass is responsible. Importantly, the reduction of rest mass also implies the elongation of measuring rods, which, in turn, implies that the accepted geometry is incorrect. Correcting for this effect reveals the true character of the geometry of space in the neighborhood of a gravitating body. In particular, the character of a black hole is quite different from that presently accepted.

Part III deals with the cosmological problem. The assumed large-scale homogeneity implies momentum conservation, which, in order to be compatible with a derived integral of motion, demands that rest masses increase in proportion to A(t), the function that is conventionally interpreted as a scale determining the increasing distance between galaxies, but which must now be identified as determining how rest masses evolve, increasing with the passage of time. It is concluded that the red shift has nothing to do with the expansion of space itself (the current understanding), but rather is a simple consequence of the fact that in the past, spectral emission frequencies were reduced because of the reduced rest masses of the era in which the emission occurred.

Part IV surveys other variable rest mass theories, and presents various detailed calculations in fourteen appendices.

Part V is a personal account of the development of the theory and of unsuccessful efforts to publish in peer-reviewed journals.

While the misconceptions surrounding special relativity are important and interesting in their own right, they also provide the key to understanding how it came about that Einstein's theory of gravity has been misunderstood and misinterpreted for decades. It is all about time, and how we understand or fail to understand its nature.

The misunderstanding regarding the nature of time began in 1905 when Einstein re-interpreted relativity theory so as to 'render superfluous' the concept of the aether, and with it the absolute nature of time and simultaneity. He did this by introducing a procedure for the synchronization of clocks that guarantees that the measured speed of light would be constant, as a tautology. By denying the existence of a preferred frame of reference, the slowing of clocks was made to seem illusory, and, all observers being equal, the concept of the relativity of time was introduced and accepted. According to this point of view, the other fellow's clocks don't run slowly for any discoverable cause, it's just that *time itself* flows more slowly in his frame of reference.

This illogical conceit appears to have influenced Einstein in his failure to fully examine the implications of his brilliant *Gedankenexperiment* presented in papers published in 1907 and 1911. Reasoning from his principle of equivalence (uniform acceleration in flat space is equivalent to being stationary in a gravitational field), Einstein predicted, among other important phenomena, the slowing of clocks in a gravitational field. Such slowing implies that all periodic phenomena, including that of atomic spectra, will suffer a reduction of frequency – a red shift. (Incredibly, many modern writers fail to recognize that the gravitational red shift is a direct consequence of the 'slowing of clocks' in a gravitational field. They 'explain' the red shift by asserting, incorrectly, that a photon must lose energy as it moves upward against the force of gravity.)

Surprisingly, Einstein failed to make any effort to discover the *cause* of clock slowing. One may speculate that he had come to embrace the relativity of time, accepting the idea that clocks are driven, so to speak, by the reduced rate of the flow of time. This attitude has become pervasive: modern writers, in referring to the gravitational slowing of clocks (= the red shift) employ the phrase '*gravitational time dilation.*'

Even more difficult to understand is the fact that Einstein also failed to explicitly recognize that his analysis also proved that light moves with constant frequency in a gravitational field. It is impossible to overstate the importance of this fact. It means that photons are not subject to the force of gravity. Thus energy and mass are not equivalent with regard to gravitational interaction. It follows that zero rest mass particles, including gravitons, do not act as sources of the gravitational field. Why did Einstein fail to follow the implications of his analysis? In this instance, it may be speculated that Einstein was perhaps reluctant to give up the universal equivalence of mass and energy.

I hope that experts in physics and mathematics will not be too put off by the (for them tedious and unnecessary) step-by step derivations presented, entailing for the most part nothing more than high school math. My purpose in taking this path is to encourage amateurs, high school students – even liberal arts majors – to follow the arguments.

Einstein, the Aether and Variable Rest Masses: Relativity Deconstructed & Reconstructed

Introduction: How relativity became mysterious

I have not taken a poll, but it is my experience that there are a great many intelligent and well-educated people who, despite real effort, feel that they do not understand relativity. Included in this group are many engineers and scientists – perhaps even a few with a PhD in physics. Their frustration has nothing to do with mathematical difficulty – relativity is very simple in that regard – but rather with an unsatisfied desire to be able to comprehend in a straightforward cause-and-effect manner the relationships between observers that relativity prescribes. In short, they want what physicists term 'a picture.' Not that there is not a surfeit of actual pictures – railway carriages and embankments, lighthouses, elevators, rocket ships, etc., appearing in many of the hundreds of textbooks and popular treatments of relativity. It is not pictures in the literal sense that are sought. These people want to know, for instance, how it can be that two observers moving past one another can each claim with perfect validity that the other guy's clocks are running slow. They want to understand how it can be, as relativists are apt to remark, that "time itself runs more slowly" in the vicinity of a massive body than it does elsewhere.

Of course, there are literally hundreds of thousands of people who are certain that they do understand relativity. This includes virtually every expert in the field. To say this is not to state the obvious – just look at quantum mechanics, of which a leading practitioner, Richard Feynman, famously said,

> "I think it is safe to say that no one understands quantum mechanics. Do not keep saying to yourself, if you can possibly avoid it, 'But how can it possibly be like that?' because you will go down the drain into a blind alley from which nobody has yet escaped. Nobody knows how it can be like that."

The difference between the experts and the unsatisfied seekers is a matter of expectation. The experts have a different attitude regarding the very possibility of having a satisfying picture of phenomena. They have a different attitude about the possibility of understanding in terms of cause and effect. In fact, they have a decidedly negative attitude regarding such expectations: they often use the adjective 'naïve' to describe such expectations.

Let me here quote Wolfgang Pauli, who in 1921, when he was only twenty-one, wrote what is still considered one of the best treatments of relativity. In this book Pauli comments on H A Lorentz's 1904 paper in which the now-famous Lorentz transformations of relativity were first presented correct to all orders of v/c. Pauli remarks,

"... even in this paper, the relativity principle was not at all apparent to Lorentz. Characteristically, and in contrast to Einstein, he tried to understand the [FitzGerald] contraction effect in a causal way."

This mindset, rejecting the necessity of understanding in terms of cause and effect, is typical of those who are satisfied with their understanding of relativity. The writers of textbooks and popularizations must follow the abstract, postulational approach employed by Einstein in his 1905 paper if they are to avoid the stigma of referring to the 'aether.'

Regarding this word: from early in the 19th century, and up until Einstein's reformulation of relativity, physicists assumed that light was a wave phenomenon that propagated at constant speed with respect to a medium, which they referred to by the old-fashioned term, aether. (The alternative spelling, ether, suggests the chemical, ethoxyethane.)

The treatment of special relativity presented here may be considered to be a variant of the Lorentz Aether Theory. However, adherents of that theory, notably Herbert Ives, have made claims that special relativity is logically inconsistent or even incorrect. Thus Ives has remarked

> "The 'principle' of the constancy of the velocity of light is not merely 'ununderstandable', it is not supported by 'objective matters of fact'; it is untenable..."

It is my view that the theory is correct but limited in scope: deductions from the theory that directly involve the notion of simultaneity cannot be trusted as representing reality. For example, the slowing of clocks: A says B's clocks run slow, but B claims that A's clocks run slow. In any other discipline, this situation would be considered to a lethal defect. But, invoking the vague idea of the relativity of time, relativists are happy to accept the reality of such a situation without offering an explanation as to how this can come about. The present treatment provides detailed explanations for this and a number of other relativity puzzles. Finally, to the best of my knowledge, no advocate of Lorentz Aether Theory has attempted a theoretical proof of the existence of the aether, as will be offered below.

Part I. The Special Theory of Relativity

The most famous assertion of special relativity is the constancy of the speed of light. This means that regardless of the motion of source or observer, the speed of light will always be measured to be a constant. Perhaps that statement doesn't sound crazy to you. But think of what it implies.

Relativity is counter intuitive: the Cosmic Flash

Imagine a large number of systems of observers moving in various directions at relativistic speeds. A flashbulb at rest in one system produces a spherical light pulse. At the instant of the flash, the location of the origin of the flash is noted in each of the systems. Subsequently these points, marked as the origin of the light pulse in each and every one of the systems, will fly apart. Einstein's basic postulate regarding the speed of light requires that observers in every system will subsequently find that the light pulse has the form of an expanding sphere that remains always centered on the marked point of origin, which is at rest in their system. How can a single sphere be centered on each of these distinct points, flying rapidly apart? This is confusing not only to laypersons, but to professionals as well. Here I quote from the introduction to Chapter 4 of Paul Richards' book, *Manual of Mathematical Physics*, regarding this situation:

> "This fact is completely foreign to intuition and cannot be reconciled with one's natural experience. The only course is to abandon intuition and investigate what this strange but inescapable *fact* implies."

Such honesty is quite rare among textbook writers. Most leave the impression that relativity's strangeness is simply an illusion resulting from our primitive preconception of the nature of space and time, and especially, of the concept of simultaneity. Einstein in his 1905 paper argued convincingly that the concept of simultaneity is an ambiguous one. Indeed, if one agrees to the very natural scheme that he suggests for the synchronization of clocks, it is clear that events judged to be simultaneous in one frame of reference will not be so in any other frame in motion with respect to the first. This circumstance is claimed to establish the 'relativity of time,' whereas, it merely indicates that there is no unambiguous way to synchronize the clocks of systems that are in relative motion. (As we shall see, the caveat – 'when only one light path connects two points' – must be added.) This undermining of the concept of time gives to relativity an air of mystery, bringing a kind of secret delight to the cognoscenti, even as it bewilders those naïve seekers of understanding, who are locked in, as it were, to the idea that light is a wave propagating in a medium, the aether. One aim of this book is to explain special relativity in a manner consistent with that primitive conception.

History of Electromagnetism and the Aether

Of course, nearly everyone has long since been convinced that the aether is a ludicrously old-fashioned concept that was forever banished by Einstein to the limbo of foolish ideas, where it joins other canards such as caloric and phlogiston. In view of this situation, it is appropriate here to make an effort to rehabilitate the reputation of the aether. This is best done by an examination of the history of progress in optics and electrodynamic theory that led up to the development of the special theory of relativity.

History does not provide us with definite beginnings, but in optics, a reasonable place to start is with the remarkable work of Thomas Young. In 1801, Young succeeded in explaining Newton's rings by supposing that light consisted of waves, which, in analogy to surface water waves, may reinforce one another or cancel one another, depending on the phases of the individual wave components. This principle of 'interference' represented a true breakthrough and may be said to mark the beginning of scientific optics. In 1817 Young made an even greater contribution. Contemplating Arago's experiment showing that two beams of polarized light do not interfere if their polarizations are at right  angles, Young developed an explanation based on the assumption that the wave vibrations were transverse to the direction of propagation.

Following Young's convincing arguments that the wave oscillations of light must be transverse rather than longitudinal, theories of a mechanical aether were faced with an extraordinary difficulty, being forced to interpret the stresses in the aether as shear, rather than compressional, stresses. The aether was thus envisioned as behaving as a rigid solid, rather than as a compressible fluid, such as air. Now real solids do in fact transmit waves of compression as well as waves of shear, but experiments in optics ruled out any compressional waves.

This circumstance represented a considerable challenge to the mathematical physicists of the day. Those involved included Cauchy, Green, Stokes, and Navier, all well-remembered as mathematical giants of the nineteenth century. Cauchy introduced a very strange theory in which the longitudinal waves were eliminated by supposing that the aether possessed a negative modulus of compression of just that magnitude, in relation to the modulus of rigidity, so as to give the longitudinal waves a zero velocity of propagation. But such an aether would be unstable with respect to contraction, as was pointed out by Green. Despite these difficulties, in 1839 James MacCullagh of Trinity College, Dublin, finally succeeded in developing a mathematical description of a highly idealized type of elastic solid whose behavior matched perfectly the results of optical experiments.
Sadly, as often happens, the scientific community somehow failed to take proper notice.

Subsequently, unaware of MacCullagh's success, others, including Gauss and his pupil, Riemann, along with Faraday, Helmholz, Wm. Thomson (later Lord Kelvin), and James Clerk Maxwell, addressed the problem of describing the aether. In 1846 Thomson put forward arguments supporting the idea that magnetism is essentially rotatory in character. (This had been the basis of MacCullagh's model.) This key idea was adopted by Maxwell, who in 1861-62 succeeded in formulating his own mechanical model of the electromagnetic field.[9]

A very remarkable feature of the model is that it implied the existence of a phenomena which had not yet been observed experimentally, namely, the so-called displacement current which occurs even in 'empty' space as a result of the time variation of the electric field. The effect was promptly confirmed, but nevertheless many scientists of the time found it difficult to accept. It took many years for Helmholz to accept the displacement current, and it seems that Lord Kelvin never really did embrace it.

Maxwell's model was quite complete and permitted him to calculate the rate of propagation of electric and magnetic effects, and the single value that he obtained was very close to that which had recently been determined experimentally by Fizeau for the velocity of light. Maxwell did not hesitate. "We can scarcely avoid the inference," he said, "that light consists of transverse undulations of the same medium which is the cause of electric and magnetic phenomena."

In 1864 Maxwell presented his views in a memoir entitled '*A Dynamical Theory of the Electromagnetic Field.*' In this great work, the theory was developed in a purely mathematical fashion, without mention of the mechanical model of the aether. In the phrase Whittaker, "the architecture of his system was displayed, stripped of the scaffolding by aid of which it had first been erected." [10]

Ironically, Maxwell was unable to successfully apply his theory to the relatively simple phenomena of optical reflection and refraction. This MacCullagh had already accomplished with his aether model and all that really needed to be done was to translate, as it were, MacCullagh's terminology to Maxwell's. The problem was not adequately treated until 1877 by Lorentz in his inaugural dissertation.

---

[9] Maxwell never believed that any mechanical model could actually represent the electromagnetic field – it was for him merely an easily investigated model from which a proper mathematical description might evolve. Banesh Hoffmann presents an accessible description of the model in his book, *Relativity and Its Roots*, W H Freeman, New York 1983, pp 69-73.

[10] E T Whittaker, *A History of the Theories of Aether and Electricity*, Vol. I, Thomas Nelson & Sons 1958, p.255

Aether and Relativity: the Michelson-Morley experiment

This mention of Lorentz brings us to the modern era, and the crisis presented by the Michelson-Morley experiment of 1887. This experiment was designed to measure the velocity of the Earth with respect to the aether. It was supposed that the apparatus was

rigid and that the apparent velocity of light in the apparatus could be determined by the usual parallelogram rule for the addition of velocities. (Incidentally, this last assumption is correct: the so-called 'relativistic composition of velocities' – Einstein's terminology – refers to the **transformation** of velocities, not to their **addition** in a single frame of reference.)

The apparatus provided two paths for light rays of equal length laid out at right angles. By means of a half-silvered mirror, a single coherent beam of light was split so that half the light was projected outward along each path. At the end of each path, a mirror reflected the half beam back to the same half-silvered mirror by means of which now the half beams were recombined to produce a series of bright and dark interference bands.

One of the paths may be taken to lie parallel to the Earth's motion through the aether. By assumption then, light traveling in that direction was reckoned to have velocity $(c - v)$, where c is the velocity of light and v the Earth's velocity relative to the aether, and velocity $(c + v)$ when traveling in the opposite direction. If L is the length of the path, the time of flight was then taken to be $L/(c - v) + L/(c + v) = 2(L/c)/(1 - v^2/c^2)$. Thus for the parallel leg, the motion through the aether was assessed to increase the time of flight by the factor $1/(1 - v^2/c^2)$. For the other leg, at right angles to the motion of the Earth through the aether, the relative velocity was the same, going and coming, and was calculated to be $c\sqrt{1 - v^2/c^2}$, according to the usual rules for combining velocities. The time of flight for the transverse leg was thus assessed to be increased, because of the Earth's motion through the aether, by the factor $1/\sqrt{1 - v^2/c^2}$, which is smaller than (being the square root of) the factor for the parallel leg, $1/(1 - v^2/c^2)$. This difference was to be detected by noting the shift of interference fringes that would occur, if all the assumptions were correct, as the apparatus was rotated. In repeated experiments the shifts observed were all very much smaller, about one-twentieth, of that expected.

Michelson himself accepted the null result as demonstrating that the aether was dragged along by the Earth. But this aether dragging idea could only be squared with the aberration of starlight, first noticed by Bradley in 1728, by means of a really tortured model evolved by Stokes and Planck. This is contrast to the beautifully simple model of aberration due to Young, in which it was assumed that the Earth has no influence upon the aether.

In 1892, Hendrik Antoon Lorentz published the first of a series of monographs outlining his Theory of Electrons. The goal of this theory was nothing less than the elucidation of all electrodynamic phenomena in terms of the interaction of moving electric charges with Maxwell's electrodynamic fields. So as to accord with Fresnel's beautiful theory of moving transparent substances, Lorentz assumed that a moving body cannot communicate its  motion to the aether that surrounds it, nor even to the aether which pervades the spaces between its particles.

The FitzGerald-Lorentz contraction

But at the time, the theory appeared to have one difficulty – it was incompetent, as originally presented, to explain the negative result of the Michelson-Morley experiment. The adjustment required was achieved by a very remarkable hypothesis: namely, that the dimensions of material bodies are altered as a result of their motion through the aether! This strange explanation was first put forward by G F FitzGerald of Trinity College sometime before June of 1889. Lorentz hit upon it independently in November of 1892.

Recall that the time of flight for a light wave moving 'upstream/downstream' was $1/\sqrt{1 - v^2/c^2}$ times the time of flight for a light wave moving cross-stream: that is, that was the computed result if the arms were assumed to be rigid and of equal length. But suppose they were not rigid, but rather, that the arm extending in the direction of the Earth's motion through the aether were to shrink by the factor $\sqrt{1 - v^2/c^2}$: this would exactly compensate for the velocity effect, explaining the null result.

The contraction hypothesis may have seemed to FitzGerald to be quite artificial. But Lorentz was able immediately to make a connection with his Electron Theory. He was able to show that any system held together by a balance of electrostatic forces must undergo a FitzGerald contraction, as it came to be called, in order to remain in equilibrium when set in motion. Lorentz realized that electrostatic forces are inadequate to describe molecular bonding, but nevertheless he felt that it was reasonable to assume that whatever forces were involved would behave similarly.[11] Incidentally, it is easy to demonstrate that the FitzGerald contraction is suffered by the Bohr atom, the name given to the simplest quantum mechanical model of the hydrogen atom, and also that its fundamental rates of oscillation would be reduced by the same factor, $\sqrt{1 - v^2/c^2}$.

---

[11] Lorentz's assumption was verified nearly 52 years later by the Hellmann-Feynman theorem.
(cf. R P Feynman, *Phys. Rev.* **56**, 340 (1939)

Any consideration of the cause of the FitzGerald contraction was made to seem irrelevant, even silly, by the appearance in 1905 of Einstein's version of the Theory of Relativity. My phrase may surprise those not familiar with the original literature, for it is part of the universal folklore that the Special Theory of Relativity was Einstein's exclusive creation. A review of the history is worthwhile.

The relativity theory of Lorentz, Larmor and Poincaré

Beginning in 1899, Lorentz began to investigate the transformation of space and time coordinates which would leave invariant Maxwell's equations. His first theory was approximate, neglecting quantities of order $(v/c)^2$ and higher.

In 1900 Joseph Larmor published transformation equations that included second order terms.[12] He was the first to show that these equations implied not only the FitzGerald contraction, but also the slowing of clock rates by the same factor. Finally, in 1904, Lorentz obtained the exact transformation, later to be named the 'Lorentz Transformation' by Poincare, and showed that Maxwell's equations for empty space were exactly invariant under it.

In 1900 Poincare suggested that electromagnetic energy might possess a mass equal to $1/c^2$ times the energy, and on September 24, 1904, in a lecture in St. Louis, Missouri, he enunciated what he called the principle of relativity – "..the laws of physical phenomena must be the same for a 'fixed' observer as for an observer who has a uniform motion relative to him." In the same lecture he declared that inertia increases with velocity, and concluded that "..there must arise an entirely new kind of dynamics which will be characterized above all by the rule, that no velocity can exceed the velocity of light."

In June of 1904, Lorentz published a paper in the proceedings of the Academy of Sciences of Amsterdam which actually confirmed Poincare's predictions and showed

further that the inertia of bodies was greater in their direction of motion than transversally. In this paper Lorentz also treated the transformation properties of Maxwell's equations in the general case with charges and currents present. In handling these sources of the fields, Lorentz had made errors, so that the complete equations appeared not to be invariant under the transformation. In June of 1905, Poincaré corrected these errors, and gave Lorentz's name to the transformation. With this paper, the Special Theory of Relativity was essentially complete – except perhaps for the matter of philosophical attitude. This brief history will make understandable the title that Sir Edmund Whittaker gave to his chapter on relativity, namely, '*The Relativity Theory of Poincaré and Lorentz.*' [13]

In any event, by June of 1905, Lorentz, Poincaré and Larmor believed that they had the correct understanding of the null result of the Michelson Morley experiment in terms of the aether, and had succeeded in developing the coordinate transformation under which Maxwell's equations, including source terms, would be invariant.

---

[12] Joseph Larmor. *Aether and Matter*, 1900, Cambridge university Press

[13] E T Whittaker *A History of the Theories of the Aether and Electricity* vol.II Thomas Nelson & Sons 1953 p.27

10

Einstein declares the aether superfluous, time relative

Now comes Einstein, who would turn the world of theory upside down. His September 1905 paper is remarkable for several reasons. Firstly, he makes no reference to the work of others: – no mention, in particular of Lorentz's papers. Various writers have asserted that Einstein at one point actually denied any knowledge of the writings of Lorentz and Poincare, and even claimed not to be aware of the Michelson Morley experiment. These claims are not credible in my view. I believe that Einstein simply wanted to avoid cluttering his paper with arguments related to the work of others. And certainly, Einstein was aware of Michelson and Morley, since in the introduction of his 1905 paper he refers to

> "..the unsuccessful attempts to discover any motion of the earth relatively to the 'light medium,'.."

Secondly, Einstein employs an approach that seems, judging from later writings, and indeed from much of the paper itself, out of character. Thus the paper adopts an approach seemingly borrowed from mathematics: it begins with two explicit but unproven postulates. The first asserts that the Galilean relativity principle is to be extended to include electrodynamics and optics. The second postulate asserts that

> "..light is always propagated in empty space with a definite velocity c which is independent of the state of motion of the emitting body."

It seems that it did not occur to Einstein that this second postulate was redundant, being an obvious consequence of the first. In fact, he reassures the reader that the second postulate is "only apparently irreconcilable with the former."

Why did Einstein, the master of the heuristic *Gedankenexperiment* employ this mathematical approach? I think that he wanted to avoid any need to deal with the aether, the erstwhile presumed medium for the transmission of light. In fact, I think that it was his intention from the outset to show that, as he remarks in the introduction of his paper

> ..the introduction of a 'luminiferous aether' will prove to be superfluous.."

Einstein's 'stealth postulate': $t_{AB} = t_{BA}$

But perhaps the strangest feature of the paper is the manner in which Einstein introduces what could be called a 'stealth postulate.' In his elaborate treatment of simultaneity and the synchronization of clocks, Einstein insists that for two co-moving points, A and B,

"..we have not defined a common 'time' for A and B unless we establish *by definition* that the 'time' required by light to travel from A to B equals the 'time' it requires to travel from B to A." [Emphasis Einstein's]

The Michelson Morley null result may be understood as proving that the speed of light, measured over a closed course, is a constant, independent of the motion of the measuring apparatus. This fact, coupled with Einstein's '*definition*,' actually makes his second postulate true as a mere tautology. This '*definition*' is actually a postulate – in fact it is *the* key postulate – upon which hangs the whole relativity program.

A single postulate for special relativity
One may usefully formulate a single postulate for special relativity as follows:

In any inertial system, the local laws of physics, including those of electrodynamics and optics, will take on some standard form, including a standard numerical content, provided that clocks are synchronized so as to insure that the measured one-way speed of light will be a constant.[14]

Thus it is clear that special relativity is limited in its range of applicability. Whenever the simultaneity of observations is involved, the theory will produce ambiguities or even seeming contradictions. The apparent symmetry regarding the slowing of clocks and the contraction of measuring rods illustrates this point. Thus system A says that B's clocks run slow, while B holds that A's clocks run slow. This ersatz symmetry has led some relativists to believe that these effects are wholly illusory.[15] An even more egregious example is the 'cosmic flashbulb' *Gedanken*, previously mentioned. A detailed treatment of this is presented in Appendix III.

Many physicists are not fully conscious of the limitations of special relativity and accept all deduction of the theory as representing reality, even those that are patently absurd. This has given scientific relativity a gloss of philosophical relativism. And this in turn has giving rise to a rather large cohort of 'cranks,' who, though not equipped to identify for themselves the limitations of the theory, are nevertheless hard-headed enough to recognize nonsense when they read it, even if it has been endorsed as gospel by a consensus of sophisticated and credentialed experts.

Incidentally, it seems that most observers credit Einstein with the development of the relativistic reformation of Newton's laws of mechanics. Thus Frank Wilczek has written,[16]

> "Einstein's great innovation was to assert the primacy of electromagnetic relativity, and work out the consequences for the rest of physics.

---

[14] Here I have shamelessly borrowed and modified R H Dicke's formulation of what he calls the Strong Equivalence Principle. cf. Dicke R H *The Theoretical Significance of Experimental Relativity* (Gordon and Breach, New York, 1965) p.4

[15] Mendel Sachs *Physics Today* 24 (1971) p. 23

[16] Frank Wilczek, 2008, *The Lightness of Being,* Basic Books, p 79.

So it was the venerable theory of Newtonian mechanics, not the upstart theory of electromagnetism that required modification."

It is true that Einstein made the assertion, but it is not true that he developed the laws of relativistic mechanics. This program was begun by Max Planck in 1906, deducing the correct form for the momentum and kinetic energy of a particle. Planck based his development on the analysis of the motion of a charged particle, leaving open the possibility that this limited the applicability of the results. In 1909, G N Lewis and Richard C Tolman, by considering the collision of two particles, verified Planck's results on the basis of the Lorentz transformation of velocities, independently of any consideration of electromagnetic theory.[17]

An otherworldly proof of aether's existence

There is a way out of this quagmire. One can prove that there is a preferred set of reference systems, at rest in the aether, with respect to which the velocity of light is truly constant. Unfortunately, because our universe is, at least for practical purposes, simply-connected and Euclidean, it is not possible for us to actually discover a preferred frame of reference. Nevertheless, the knowledge that the aether exists is of paramount importance in making relativity comprehensible, and in understanding the limitations of the theory.

The special theory of relativity is a local theory. All that is required is that the geometry of the region of interest must be Euclidean. There is no restriction regarding the topology of the entire space. Thus, special relativity would be valid in a universe possessing the topology of a flat three-torus, a space which is locally Euclidean everywhere.

The three-torus is easily imagined in terms of a box and a simple set of rules that identify points on the boundary of the box. Points diametrically opposite one another on the left-right, front-back, top-bottom surfaces are declared to be the same point. In this space, there is a multiplicity (actually, an infinity) of geodesic (locally minimum-length paths) connecting any two distinct points in the space. Certain paths are of special interest. These are those whose three 'winding numbers' are all zero. The winding number for the x-axis (the normal to the left-right surfaces) is the length of the path in the x-direction divided by the diameter of the box in the same x-direction, rounded down to an integer. There is first of all a special path that goes directly from one point, A, to the other, B, without passing through any of the bounding surfaces. In addition there are three indirect paths of winding number zero that penetrate one of the bounding surfaces just once. So there are four such special paths altogether. Actually, it is much easier to visualize these paths if first one imagines that the box is repeated by parallel displacement in all three directions (forming a three-dimensional analog of wallpaper, so to speak). Then the indirect paths are simply straight lines connecting point A in the original box with the image, B', of point B in one of the six boxes immediately adjacent. Of these paths, only three will have a zero winding number, and these are the ones that need to be investigated.

---

[17] Sir Edmund Whittaker, *A History of the Theories of Aether and Electricity, Vol. II*, pp. 44-51.

As I have argued, the key postulate in the special theory of relativity is the so-called '*definition*' asserting that, for commoving points A and B, the time required for light to go from A to B is the same as the time for light to go from B to A. This assumption forms the basis of the scheme for synchronizing clocks and defining simultaneity in a system. In an infinite Euclidean space there is only one geodesic path connecting two points, but in the three-torus there are four 'shortest' paths. So when one carries out Einstein's synchronization procedure along each of the four paths connecting A and B, the result for the setting of B's clock will either be consistent, or not. Suppose that for some system of observers, the procedure yields consistent results. Call this an aether system. I leave it as an exercise for the experts to prove that for any system in motion with respect to an aether system, the synchronization procedure will fail to yield consistent results.

But the situation may be understood by simply observing that if the direct path from A to B has a component parallel to the velocity of the system with respect to an aether system, then $t_{AB} > t_{BA}$. Thus the final signal sent to synchronize the clock at B will take longer to arrive than the average time, $(t_{AB} + t_{BA})/2$, which is assumed in Einstein's synchronization procedure. This will leave the clock at B retarded relative to A's clock. On the other hand, the indirect path from A to B′ will be more or less anti-parallel to the motion of the system, implying that $t_{AB'} < t_{B'A}$, and in this case the synchronizing signal will arrive 'too soon,' making the clock at B advanced relative to A's clock.

Finally, it is unimaginable that a preferred system, at rest in the aether, exists only in closed spaces such as the three-torus: it must be a feature of all spaces regardless of topology. So the naïve concept that light is a wave phenomenon moving with constant speed in a medium, the aether, is correct. Our objective in the following sections will be to show that this view is compatible with the theory of relativity, and indeed, is the only way to make sense of that theory.

Incidentally, it is my opinion that the obvious dipole structure in the spatial distribution of the cosmic microwave background radiation constitutes a proof of the existence of the aether. Relativists were curiously silent regarding this, quietly assenting to the obvious interpretation that the Earth's motion was responsible. None, however, attempted to explain why in this instance Lorentz invariance did not apply.

Also, the existence of the aether provides an answer to a vexing problem which arises in connection with Alain Aspect's brilliant experiment testing Bell's theorem. The results demonstrated conclusively that a nonlocal (faster than light) influence was involved. Bell was worried that such faster than light influence would imply, according to special relativity, the possibility of going backward in time in other inertial reference frames. Bell suggested that an answer to this problem was to return to the Lorentzian relativity, accepting the reality of a preferred frame of reference, the aether, in which causal sequences are unambiguous.

Regarding coordinate systems and systems of observers

In special relativity, a system of observers is defined so that, at any event of interest in spacetime, there will be, at that event, an observer co-moving with the system. The observer is considered to be equipped with a real clock that has been synchronized according to the 'equal-times' method prescribed by Einstein. Such an observer can thus report the 'what, where and when' for the event of interest.

In special relativity the coordinate intervals correspond directly to measurements made using real measuring rods and real clocks. To observers in the system, the rods are seemingly rigid, and the clocks appear to keep perfect time, but if the system is not at rest in the aether, neither is true: rods will be contracted in the direction of the system's motion through the aether, and clocks will run more slowly than clocks at rest in the aether. For the present, the reader is asked to accept these phenomena – contraction and slowing – as postulates: later I will present arguments in support of their reality.

The principle systems of interest are, naturally, a system, K, at rest in the aether and a 'primed' system, K' that is in motion with respect to K. (For simplicity, the motion is usually imagined to take place in the $+x = +x'$ direction.) It will often be convenient to consider an imaginary system, K*, that moves with K', but does not suffer the contraction of measuring rods or the slowing of clocks. The time in K* is 'borrowed' from the aether system, K, and distance intervals measured in K* are the same as measured in K. Thus there will be no need to introduce coordinates with asterisks appended.

The Lorentz Transformation & the 'Wave of Simultaneity'

Although the derivation of the Lorentz Transformation is presented in most of the great many texts on the subject of relativity, we repeat the process here in order to show how it can be derived (and understood!) from the aetherist standpoint. As a preliminary, we will introduce the concept of the wave of simultaneity.

Recall Einstein's 'definition' – his stealth postulate – asserting that "..the 'time' required by light to travel from A to B equals the 'time' it requires to travel from B to A." In the K (aether) system, the speed of light is uniform in all directions. As assessed in the K system, light moves relative to the K' system with velocity $c - v$ in the x direction and $c + v$ in the $-x$ direction. This means that observers of the K system will not agree that the clocks in K' are synchronized: if observers in the K' system emit little flashes of light 'simultaneously,' in their opinion, observers in K will see a 'wave of simultaneity' moving rapidly in the direction in which K' itself moves.

To compute the speed, W_s^*, of this 'wave of simultaneity,' it is convenient to adopt the imaginary system K*, in which the aether appears to stream in the $-x$ direction with speed, v. (The measuring rods and clocks in K* are assumed to be unaffected by their motion with respect to the aether, and the time is 'borrowed' from the K system.) Observers in K′, the moving system, synchronize clocks assuming that signals propagate at the speed of light in all directions. In synchronizing a clock that is 'upstream' in the aether wind, it is assumed that the start signal will arrive after the time interval, here expressed in terms of measurements in K*,

$$\frac{1}{2}(t_{AB} + t_{BA}) = \frac{1}{2}\left[\frac{D}{c-v} + \frac{D}{c+v}\right] = \left[\frac{1}{1-v^2/c^2}\right]\frac{D}{c},$$

But in fact the signal in fact requires $t_{AB} = \left[\frac{1}{1-v/c}\right]\frac{D}{c}$ and consequently, the

signal arrives late by $\Delta t = t_{AB} - \frac{1}{2}(t_{AB} + t_{BA}) = \frac{1}{2}(t_{AB} - t_{BA}) = \left[\frac{v/c}{1-v^2/c^2}\right]\frac{D}{c}$. Then

$$W_S^* = \frac{D}{\Delta t} = \frac{c^2}{v}(1 - v^2/c^2) = \frac{c^2}{v} - v$$

Incidentally, since K* moves with velocity, v, in the $+x$ direction with respect to the aether system, K, the wave of simultaneity will move relative to K at speed

$$W_S = W_S^* + v = \frac{c^2}{v}$$

An aetherist derivation of the Lorentz transformation

In deducing the Lorentz transformation, we shall use only the above result for the wave of simultaneity, together with the two real physical effects: the FitzGerald contraction of measuring rods and the Larmor slowing of clocks, lengths and rates, each being reduced by the factor $\sqrt{1-v^2/c^2} = \gamma^{-1}$.

The transformation of position is simplicity itself. An event located at x, t will be judged by K^* to be located at $(x - vt)$, while observers in K′ , because of their shrunken measuring rods, will judge it to be located at

$$x' = \gamma(x - vt) = \gamma(x - \beta ct)$$

The situation for the time transformation is more complicated. First, at time t the origin of the K′ system will be located at $x_0 = vt$, and the clock there will read, because of its slowing, $\gamma^{-1}t$. The clock of interest, further on at $x > x_0$ (also observed at time, t) will be retarded relative to the clock at x_0 by the amount

$$\left[\frac{x - x_0}{W_S^*}\right]\gamma^{-1} = \frac{x - vt}{(c^2/v)(1 - \beta^2)} = \gamma\left[\frac{vx}{c^2} - \frac{v^2 t}{c^2}\right]$$

The clock at x will therefore read

$$t' = \gamma^{-1}t - \gamma\left[\frac{vx}{c^2} - \frac{v^2}{c^2}\right] = \gamma\left[\gamma^{-2}t + \frac{v^2t}{c^2} - \frac{vx}{c^2}\right] = \gamma\left[t - \frac{vx}{c^2}\right] = \gamma(t - \beta ct)$$

So finally we have

$$x' = \gamma(x - \beta ct) \quad ct' = \gamma(ct - \beta x)$$

Thus we have shown that the Lorentz transformation is a simple consequence of the demand that the speed of light be made constant by the artifice of adjusting the synchronization of clocks, together with the physically real effects – the shortening of measuring rods and the slowing of clocks.

The transformation (not the addition) of velocities

It is to be emphasized that the usual 'parallelogram' rule for *adding* velocities relative to a single coordinate system remains valid in relativity. Unfortunately, many writers confuse the issue by using the phrase 'the composition of velocities,' or worse, 'the addition of velocities' when referring to what is properly called 'the *transformation* of velocities.'

We consider three systems: K, an aether rest system, K', which moves relative to K in the $+x$ in direction with uniform speed v, and K*, which moves with K' but suffers neither a shortening of measuring rods nor a slowing of clocks.

Event 1: O and O', origins of K and K', coincide at time $t = t' = 0$

Event 2: a moving object is detected at x at time t, where $x = ut$ while O′ is at

$$x_{o'} = vt$$

At the same time (according K) the moving object is detected at P' in K'. The distance, $d*$, from O' to P' is measured in K* as

$$d* = ut - vt, \quad \text{whence observers in K' will measure this as}$$

$$d' = \gamma\, d* = \gamma\,(u - v)t$$

The time of event 2 according to O' is $t' = \gamma^{-1}t$, and the time of this event according to the observer at P' is

$$t'_{P'} = t' - \gamma^{-1}(d*/W_S{}*)\,,$$

where $W_s^* = c^2/v - v$ is the speed of the wave of simultaneity relative to K*.

$$t'_{p'} = \gamma^{-1}[1 - \frac{u-v}{c^2/v-v}]t = \gamma^{-1}[(c^2/v-v)-(u-v)]/(c^2/v-v)$$

$$= \gamma^{-1}\frac{c^2-uv}{v}\frac{v}{c^2-v^2} = \gamma^{-1}\frac{1-uv/c^2}{1-v^2/c^2} = \gamma(1-uv/c^2)$$

According to K', the object has covered the distance

$$d' = \gamma(u-v)t \quad \text{in time} \quad t'_{p'} = \gamma(1-uv/c^2)t \quad \text{and the speed is thus}$$

$$u' = \frac{u-v}{1-uv/c^2}$$

A formal derivation of the Lorentz transformation, emphasizing the logic involved, is presented in Appendix I.

Paradox Lost

Having deduced the Lorentz transformation, it would be correct to say that the seeming paradox regarding the length of measuring rods and the rate of clocks has been resolved since the transformation and its inverse exhibit the same reciprocity as do the measurements in question. But it will be worthwhile to show how the synchronization convention together with the real physical effects of length contraction and clock slowing conspire to produce the seeming paradoxical results.

The Measuring Rod 'Paradox'

Here we seek to explain how measurements made by observers in a moving frame can claim that a measuring rod in the rest frame is shortened – this in spite of the fact that it is *their* measuring rods that are actually shortened. We shall use the previously derived expression for the speed of the 'wave of simultaneity,' namely

$$W_S^* = \frac{c^2}{v}(1-v^2/c^2) = \frac{c^2}{v} - v$$

Consider the measurement of a rod of length, L, at rest in the K (aether) frame by observers in the moving K' frame. After the left (downstream) end is observed, the synchronization signal zooms off upstream at speed, W_s, while the right end of the rod moves to the left at the leisurely speed, v. These will meet at the observer who thinks he is making his observation of the right end of the rod simultaneously with the observation of the left end, which was actually made earlier. The time, δt, (as assessed in the K* system) required for this meeting satisfies

$$W_S^* \delta t + v \delta t = L \quad \Rightarrow \quad \delta t = \frac{L}{W_S^* + v} = (v/c)\frac{L}{c}$$

The measured length of the rod, in terms of aether frame units, may be written as

$$L* = L - v\delta t = (1 - v^2/c^2)L$$

But observers in K' exaggerate distances because of their shortened measuring rods, and so they get

$$L' = L*/\sqrt{1 - v^2/c^2} = \sqrt{1 - v^2/c^2}\,L$$

The Clock 'Paradox'

To represent a clock in the K (aether) system we imagine a beam of length L set at right angles to the direction of motion of the K' system which moves to the right with speed v. At the outer end of the beam there is a mirror, and the time is that required for light to traverse the course, out and back, is $T = 2L/c$. In that time, the inner end of the beam will have moved to the left (relative to K*) the distance, $D = vT$.

When the K' observer, A, sees the pulse of light launched, the observer to the left, B, who will eventually see the pulse return to the base of the beam, will have a clock reading that is advanced relative to that of observer, A. We have seen that the speed of the 'wave of simultaneity' is $W_S^* = \frac{c^2}{v}(1 - v^2/c^2)$, so the clock at B is advanced, in terms of 'true' (aether) time, by

$$\frac{D}{W_S^*} = \left[\frac{v/c}{1 - v^2/c^2}\right]\frac{D}{c}$$

But since clocks run slow by the factor $\sqrt{1 - v^2/c^2}$ in K', the clock at B at the instant of launch of the pulse is actually advanced by

$$\gamma^{-1}\frac{D}{W_S^*} = \left[\frac{v/c}{\sqrt{1 - v^2/c^2}}\right]\frac{D}{c}$$

In system K, the pulse requires time $T = D/v$ to complete return to the base, but again in system K', the clocks will have advanced by only $\sqrt{1 - v^2/c^2}\,\frac{D}{v}$. The reading of B's clock will then be advanced relative to A's by the amount

$$\sqrt{1 - v^2/c^2}\,\frac{D}{v} + \left[\frac{v/c}{\sqrt{1 - v^2/c^2}}\right]\frac{D}{c} = \frac{1}{\sqrt{1 - v^2/c^2}}\frac{D}{v} = \frac{1}{\sqrt{1 - v^2/c^2}}T$$

Thus the period of the clock at rest in K will appear to observers in K' to be longer by the factor γ.

It should be clear that deductions from the theory that directly depend upon the synchronization of clocks are not to be trusted. But it must be emphasized that this does not indicate that the theory is incorrect: one just need bear in mind this limitation.

Why do measuring rods contract and clocks slow?

One may hope that, after the foregoing demonstrations, most readers will concede that the seeming paradoxes of relativity can be understood within an aether framework, provided that one can accept the reality of the FitzGerald contraction of measuring rods, and the Larmor slowing of clock rates. I previously mentioned Lorentz's argument that the contraction effect was to be expected as a consequence of the transformation laws for the electric field (which he was the first to correctly describe). In the modern era, John Stewart Bell used the same transformation law along with equations of relativistic mechanics to show that a slowly accelerated atom will suffer the FitzGerald contraction and the Larmor slowing.[18] But these arguments are in a certain sense circular, relying as they do on relativity theory. Fortunately, it turns out to be possible to deduce these effects by means of an analysis which simply recognizes the wave nature of all matter.

Harter's standing wave model for matter

Prior to the development of quantum mechanics, material objects were viewed as 'hard things' existing in the aether, which filled all space. Physicists puzzled as to how material objects could move freely through the aether, which transmitted transverse waves in the manner of a rigid solid. When Einstein declared the concept of the aether to be 'superfluous,' most physicists were delighted. Good riddance! Matter could now be pictured as 'hard things' existing in empty space. But when, with the advent of quantum theory, the wave nature of all matter was revealed, another point of view became possible – namely, that all material objects might be characterized as wavelike disturbances in a medium, the aether.

The 'Standing Wave Model' to be presented was developed by Professor William G Harter of the University of Arkansas.[19] It is admittedly a 'toy model' and is only intended to show the feasibility of a wave characterization of matter. Quite surprisingly, the model not only correctly

[18] J S Bell, *Speakable and unspeakable in quantum mechanics,* 1987, Cambridge University Press, pp.67-80.

[19] W. G. Harter, J. Mol. Spectrosc. 210, 166(2001).
W. G. Harter, J. Evans, R. Vega, and S. Wilson, Am. J. Phys. 53, 671(1985).
W.G. Harter and T. C. Reimer, www.uark.edu/ua/pirelli/html/poincare_inv_2.htm
(It must be said that Professor Harter does not accept the literal existence of the aether.)

predicts the FitzGerald contraction and the Larmor frequency reduction; it also yields the relativistic expressions for energy and momentum!

Three assumptions are made. First, the constituent waves propagate at the speed of light independently of frequency (no dispersion). Second, observers moving with the object will see it as a standing wave, while observers at rest in the aether will see a standing wave configuration in uniform motion. Third, in order to deduce the expressions for energy and momentum, we assume that the usual quantum relation, $E = \hbar\omega$, between the energy, E, and frequency, ω, holds.

Extended objects

We first consider an object such as an atom, molecule, crystal, or perhaps a measuring rod, that has spatial extension, in contrast to point particles, which will be considered later. The ends of the object define the limits of the waves and may be thought of as acting as mirrors, and the line of motion is taken to be parallel to line joining the ends.

In the frame co-moving with the object, the standing wave has the form

$$\Psi' = \exp(-i\omega_0 t')\cos k_0 x'$$

In the frame of the aether, which moves with speed, βc, in the minus x direction with respect to the object, the wave function comprises the superposition of two wave trains, one moving in the $+x$ direction with wave number $k_\rightarrow$ and frequency $\omega_\rightarrow$, and one moving oppositely with wave number $k_\leftarrow$ and frequency $\omega_\leftarrow$:

$$\Psi = \exp[i(k_\rightarrow x - \omega_\rightarrow t)] + \exp[i(k_\leftarrow x - \omega_\leftarrow t)]$$

Making use of the mathematical identity,

$$\exp(ia) + \exp(ib) = \exp[i(a+b)/2]\{\exp[i(a-b)/2] + \exp[-i(a-b)/2]\}$$

$$= 2\exp[i(a+b)/2]\cos[(a-b)/2]$$

we can write

$$\Psi = 2\exp\{ i[(k_\rightarrow + k_\leftarrow)/2]x - i[\omega_\rightarrow + \omega_\leftarrow)/2]t\} \otimes$$
$$\cos\{[(k_\rightarrow - k_\leftarrow)/2]x - [(\omega_\rightarrow - \omega_\leftarrow)/2]t\}$$

Now $\omega_\rightarrow$ will be blue-shifted relative to ω_0: $\omega_\rightarrow = b\omega_0$, $b \geq 1$, to be determined, and $\omega_\leftarrow$ will be red-shifted by the reciprocal factor: $\omega_\leftarrow = b^{-1}\omega_0$. Also, $k_\rightarrow = bk_0$, and $k_\leftarrow = b^{-1}k_0$. Thus we can write

$$\Psi = 2\exp\{ i[(b-b^{-1})/2]k_0 x - i[(b+b^{-1})/2]\omega_0 t\} \otimes$$
$$\cos\{[(b+b^{-1})/2]k_0 x - [(b-b^{-1})/2]\omega_0 t\}$$

Now the cosine factor represents the envelope and hence the group behavior of the wave. Thus, using $\omega_0 / k_0 = c$,

$$x = [(b - b^{-1})/2]\omega_0 t / [(b + b^{-1})/2]k_0 = \{[b^2 - 1]/[b^2 + 1]\}ct$$

Thus $\beta = [b^2 - 1]/[b^2 + 1]$ is the relativity parameter of the object with respect to the aether frame. Solving for b we find

$$b = \sqrt{(1 + \beta)/(1 - \beta)},$$

the well-known Doppler shift factor. Also,

$$[b + b^{-1}]/2 = 1/\sqrt{1 - \beta^2} \quad \text{and} \quad [b - b^{-1}]/2 = \beta/\sqrt{1 - \beta^2}$$

Inserting these we can write

$$\Psi = 2\exp\{i[(\beta/\sqrt{1 - \beta^2})k_0 x - (1/\sqrt{1 - \beta^2})\omega_0 t]\} \otimes$$
$$\cos[(1/\sqrt{1 - \beta^2})k_0 x - (\beta/\sqrt{1 - \beta^2})\omega_0 t]$$

The overall length of the standing wave as assessed in the rest (aether) frame is given by considering the cosine (envelope) factor at a fixed time. If x_0 and x_N represent the ends of the standing wave, which comprises N full cycles, we have

$$x_N - x_0 = 2\pi N \sqrt{1 - \beta^2} / k_0$$

This demonstrates the FitzGerald contraction.

Regarding the rate of the on-board 'clock,' we must look at the dynamic factor (the exponential) at a fixed location in the standing wave. Thus in the exponential we set $x = \beta ct = \beta(\omega_0 / k_0)t$, and determine the time required for one complete cycle. The exponential takes the form

$$\exp\{i[(\beta/\sqrt{1 - \beta^2})k_0(\beta(\omega_0 / k_0)t) - (1/\sqrt{1 - \beta^2})\omega_0 t]\} = \exp\{-i\sqrt{1 - \beta^2}\,\omega_0 t\}$$

Then if t_0 and t_1 are the times for the start and finish of one cycle,

$$t_1 - t_0 = (2\pi / \omega_0)/\sqrt{1 - \beta^2}$$

showing that the Larmor slowing is inherent in the standing wave picture of matter. But one can go further, deducing the fundamental relativistic expressions for energy and momentum. According to quantum theory, energy and momentum are related to frequency and wave number by the relations $E = h\omega$ and $p = hk$ respectively. Thus from the final expression for Ψ, we find for the energy and momentum

$$E = h\omega = h\omega_0[1/\sqrt{1-\beta^2}] \quad \text{and}$$

$$p = hk = hk_0 [\beta/\sqrt{1-\beta^2}] = h\omega_0/c [\beta/\sqrt{1-\beta^2}]$$

Then, if we may identify $h\omega_0$ as the rest energy mc^2, we may write

$$E = mc^2/\sqrt{1-\beta^2} \quad \text{and} \quad p = mc\beta/\sqrt{1-\beta^2}$$

Thus, as if by magic, the representation of matter as a standing wave produces not just the FitzGerald contraction and the Larmor slowing, but the correct relativistic expressions for energy and momentum.

The reader may have noticed that the change in amplitude on reflection has been ignored in the analysis presented above. It turns out that this involves a rather tedious calculation, presented in appendix XIV, which, however, has no effect on the conclusions presented above.

Point particles
Here I have adapted Harter's approach to enable the modeling of 'point particles,' such as the electron. The aim is to show that such particles may be modeled as resonances of waveguide-like structures in a presumed rolled-up extra dimension. I consider here a standing wave system in which the waves propagate in a direction, w, which is at a right angle to any possible direction of motion in ordinary space, as well as moving in ordinary space in the +x direction. In the frame moving with the particle one has

$$\Psi' = 2\exp(-i\omega_0 t') \cos(k_0 w')$$

I use w rather than y or z to indicate that this dimension is not in our usual space.

In the frame of the medium, the aether, which moves with speed, βc, in the minus x-direction with respect to the particle, the wave function is

$$\Psi = \exp[i(k_{\uparrow x}x + k_{\uparrow w}w - \omega t)] + \exp[i(k_{\downarrow x}x + k_{\downarrow w}w - \omega t)]$$

Here we have written the wave-number vectors as $\vec{k}_\uparrow = k_{\uparrow x}\hat{x} + k_{\uparrow w}\hat{w}$ and $\vec{k}_\downarrow = k_{\downarrow x}\hat{x} + k_{\downarrow w}\hat{w}$, where the up arrow ($\uparrow$) indicates propagation in the $+w$ direction, and ($\downarrow$) in the $-w$ direction. In order that the wave will hang together, so to speak, we must have

$$k_{\uparrow x} = k_{\downarrow x} \equiv k_x \text{ and } k_{\uparrow w} = -k_{\downarrow w} \equiv k_w$$

Then, again using the identity

$$\exp(ia) + \exp(ib) = \exp[i(a+b)/2]\{\exp[i(a-b)/2] + \exp[-i(a-b)/2]\}$$

$$= 2\exp[i(a+b)/2]\cos[(a-b)/2],$$

we may write for the wave

$$\Psi = 2\exp[i(k_x x - \omega t)]\cos(k_w w)$$

Also, the cosine factor above must fit the same term in the first equation, so

$$k_w = k_0$$

Introduce factors, a and b, to be determined, setting $k_x = ak_0$ and $\omega = b\omega_0$
Since one has always $\omega = c|k|$, we have

$$b\omega_0 = c\sqrt{k_x^2 + k_w^2} = ck_0\sqrt{a^2 + 1} \implies b^2 = a^2 + 1$$

We'll use this directly. From the last equation for Ψ, the phase velocity in the x-direction is

$$V_{ph} = \omega/k_x = (b/a)(\omega_0/k_0) = (b/a)c$$

We must insure that the group velocity, V_{gr}, matches the velocity of the particle, βc, in the x direction. Now the group velocity is defined by $V_{gr} \equiv \partial\omega/\partial k$. Noting that k_0 is a constant, we have

$$\partial\omega/\partial k = \partial/\partial k_x(c\sqrt{k_x^2 + k_0^2}) = ck_x/\sqrt{k_x^2 + k_0^2}$$
$$= c^2 k_x/\omega = c^2(a/b)(k_0/\omega_0) = (a/b)c$$

Thus, $V_{gr} = \beta c \implies (a/b)c = \beta c \implies a = \beta b$

Combining this with $b^2 = a^2 + 1$ one has

$$b^2 = b^2\beta^2 + 1 \qquad \Rightarrow \qquad b = 1/\sqrt{1-\beta^2} \quad \text{and} \quad a = \beta/\sqrt{1-\beta^2}$$

Note that $\omega = b\omega_0 = \omega_0\sqrt{1+a^2} = \omega_0\sqrt{1+(k_x^2/k_0^2)} = \sqrt{\omega_0^2 + c^2 k_x^2}$

Thus it is just the momentum in the direction of motion in ordinary space that enters.

Finally, we have, assuming $h\omega_0 = mc^2$ and $hk_x = h(ak_0) = h(\omega_0/c)a = mca$

$$E = h\omega = h\omega_0 b = mc^2/\sqrt{1-\beta^2} \quad \text{and} \quad p = hk_x = mca = mc\beta/\sqrt{1-\beta^2}$$

These are the standard relativistic expressions for energy and momentum.

Regarding the slowing of the 'on-board clock,' aka, 'time dilation': Going back to the equation for the wave, namely, $\Psi = 2\exp[i(k_x x - \omega t)]\cos(k_w w)$, we look at the wave at a fixed location in the co-moving frame, i.e., for $x = \beta ct$. Inserting this along with the relations $k_x = ak_0$, $k_w = k_0$ and $\omega = b\omega_0$, we have

$$\Psi = 2\exp[i(ak_0\beta ct - b\omega_0 t)]\cos(k_0 w) = 2\exp[i(a\beta\omega_0 t - b\omega_0 t)]\cos(k_0 w)$$

$$= 2\exp[i(a\beta - b)\omega_0 t]\cos(k_0 w)$$

$$= 2\exp[i(\beta^2/\sqrt{1-\beta^2} - 1/\sqrt{1-\beta^2})\omega_0 t]\cos(k_0 t)$$

$$= 2\exp[-i\sqrt{1-\beta^2}\,\omega_0 t]\cos(k_0 w)$$

This demonstrates the frequency reduction effect. Needless to say, determining the contraction of a point particle is out of the question.

I hope that the foregoing will satisfy those who might have been skeptical of the reality of the FitzGerald contraction and the Larmor reduction of frequencies.

Incidentally, this standing wave model brings to mind a remarkable vision expressed by Michelson in his one of his 1899 Lowell lectures:

> "..[this is] one of the grandest generalizations of modern science, namely, that all the phenomena of the physical universe are only different manifestations of the various modes of motion of one all-pervading substance – the aether."

It may also be remarked that Heisenberg's Uncertainty Principle is just a simple mathematical consequence of the wave nature of matter.

Two final exercises in Special Relativity

An alternate method for synchronization: Slow transport of clocks

The reader may feel that Einstein's stealthy postulate – his prescription for synchronizing clocks – has at this point been quite thoroughly examined, and has been found to be the key to understanding the seeming paradoxes of Special Relativity. But perhaps there may be some alternative method for the synchronization of clocks. One method that has been suggested is to synchronize all of the clocks at one point, and then slowly carry the clocks to their designated stations. We shall see that this procedure leads to the same result as the Einstein process.

We consider carrying a clock from the location of the master clock at x_m, y_m, z_m to its appointed station at x_s, y_s, z_s. Clocks in the K' system run slow as compared with those in the aether system, K: $dt' = \sqrt{1 - v^2/c^2}\, dt$. During the period of transport, v is taken to be constant. The difference in the readings of the master and the slave clock will be

$$t'_m - t'_s = \int_{t_m}^{t_s} \sqrt{1 - v^2/c^2}\, dt - \int \sqrt{1 - [(v + dx/dt)^2 + (dy/dt)^2 + (dz/dt)^2]/c^2}\, dt$$

To first order in the transport speed, one has

$$t'_m - t'_s \approx \int_{t_m}^{t_s} \sqrt{1 - v^2/c^2}\, dt - \int_{t_m}^{t_s} \sqrt{1 - [v^2 + 2v(dx/dt)]c^2}\, dt$$

$$= \int_{t_m}^{t_s} \sqrt{1 - v^2/c^2}\, dt - \int_{t_m}^{t_s} \sqrt{1 - v^2/c^2 - 2v(dx/dt)c^2}\, dt$$

$$= \int_{t_m}^{t_s} \sqrt{1 - v^2/c^2}\, dt - \int_{t_m}^{t_s} \sqrt{1 - v^2/c^2} \sqrt{1 - \frac{2v(dx/dt)}{(1 - v^2/c^2)c^2}}\, dt$$

$$\approx \int_{t_m}^{t_s} \sqrt{1 - v^2/c^2}\, dt - \int_{t_m}^{t_s} \sqrt{1 - v^2/c^2} \left[1 - \frac{v(dx/dt)}{(1 - v^2/c^2)c^2} \right] dt$$

$$= \frac{v}{c^2 \sqrt{1 - v^2/c^2}} \int_{t_m}^{t_s} (dx/dt)\, dt = \frac{v}{c^2 \sqrt{1 - v^2/c^2}} \int_{x_m}^{x_s} dx$$

$$= \frac{v}{c^2 \sqrt{1 - v^2/c^2}} [x_s - x_m]$$

This is the difference in the readings of the proper clocks in the moving system, as observed in K, the aether system. If each of these clocks emits a signal 'simultaneously,' the time interval (measured in K) between detection of the signal from the master clock and that from slave clock will be larger that the difference in the readings by the factor

$1/\sqrt{1 - v^2/c^2}$, and thus will be given by

$$t_m - t_s = \frac{v}{c^2(1 - v^2/c^2)}[x_s - x_m]$$

In the hypothetical system, K*, which moves with K', but is immune to the contraction and slowing effects, the wave of simultaneity would appear to move with speed

$$W_S^* = \frac{x_s - x_m}{t_m - t_s} = \frac{c^2}{v}(1 - v^2/c^2)$$

exactly as previously derived assuming Einstein's method of synchronization.

Bell's Rockets: An argument supporting aether's existence

John Bell, in his remarkable collection, *Speakable and unspeakable in quantum mechanics*, includes an essay titled 'How to teach special relativity.' Bell offers a simple thought experiment to illustrate the fact that standard presentations of the subject all too often "..destroy completely perfectly sound and useful concepts already acquired."

Bell imagines identical rockets, one leading the other, undergoing identical acceleration programs. He further imagines that they are tied together with a taut but fragile thread. The question: will the thread break? Bell relates that the question was put to the members of the CERN Theory Division, and a clear consensus insisted that the thread would *not* break (Bell's emphasis).   Of course, after careful calculation, everyone finally agreed that the thread would break. But as Bell points out, calculations are unnecessary: one only has to consider the situation from the standpoint of observers on the launch pad, and accept the reality of the FitzGerald contraction. Actually, it's a bit more complicated, and I believe that it will be useful to consider this little problem in greater detail from the aetherist standpoint.

Imagine two rockets of identical design, one behind the other separated by nominal distance D (measured by proper observers in their frame of reference) along intended direction of motion. An observer midway between the rockets simultaneously receives signals from the pair of rocketeers indicating that their clocks are synchronized. The observer immediately sends an order to blast-off. To keep things simple, we imagine that the acceleration is instantaneous, so on receipt of the signal, the rockets 'jump' to velocity, v. We also ignore rocket exhaust and other non-essentials. Measurements in the aether frame K will not carry primes; those in the rocket system K' will carry a prime; measurements made before blast off will be subscripted with the numeral, 1, those after with numeral, 2.

We first assume that the rockets were initially at rest in an aether rest frame, implying that $D_1 = D_1' = D$. According to aether observers, the rockets are at all times separated by the same distance $D_2 = D_1 = D$, and according to these observers, the rocketeer's clocks remain synchronized. However, the rates of the clocks will be judged to be reduced by the factor. Also, because of the FitzGerald contraction effect, the rockets and everything on board, including measuring rods, will be shortened in the direction of motion by the factor $\sqrt{1 - v^2/c^2}$. Thus the rocketeers will, with their shortened measuring rods, measure the distance between the rockets to be

$$D_2' = D_2 / \sqrt{1 - v^2/c^2} = D / \sqrt{1 - v^2/c^2}$$

Also, the rocketeers will discover that their clocks appear to be no longer synchronized. To determine this, they proceed as follows. They agree that when their clocks have a certain reading, they will each send a signal to a co-moving cohort stationed midway between the rockets. According to aether observers, the signal from the trailing rocket will move with velocity $c - v$ toward the cohort, while that from the lead rocket will move with speed $c + v$.

The time interval between the arrival of the signals from the rockets will be, according to the aether observers

$$\Delta T = D[1/(c-v) - 1/(c+v)]/2 = (Dv/c^2)/[1 - v^2/c^2]$$

But according to co-mover cohort, whose clock runs slow by the factor $\sqrt{1 - v^2/c^2}$, the difference is

$$\Delta T_2' = \sqrt{1 - v^2/c^2}(Dv/c^2)/[1 - v^2/c^2] = (Dv/c^2)/\sqrt{1 - v^2/c^2}$$

The point is this: if it were true that 'everything is relative', one would expect that two identical machines (the rockets), undergoing identical acceleration episodes would end up in a state indistinguishable from the initial state. Since this is not the case, it seems logical to accept that there must be something 'out there' (the aether) that causes the observed changes.

But what if we imagine that initially the rockets are moving backwards through the aether and then 'jump' so as to come to rest in the aether? In this case, the synchronization technique plays a crucial role. In this situation, $D_1' = D$, and $D_1 = D\sqrt{1 - v^2/c^2}$. We are here accepting the proper length, D – that measured by local observers – as the given: it is not equal to the D of the previous situation.

If at time t_1 the midpoint observer receives the timing signals sent from the two rockets, 'simultaneously' at clock time t_0, the aether observers will know that the signal must have been sent by the 'lead' rocket at time $t_1 - (1/2)D_1/(c - v)$, so at time t_1, the 'lead' rocket's clock, which runs slow by the factor $\sqrt{1 - v^2/c^2}$, will read

$$t_0 + (1/2)\sqrt{1 - v^2/c^2}\, D_1/(c - v)$$

We may assume that the midpoint observer sent the signal to blast off immediately at time t_1. If so, it will arrive at the lead rocket after an interval of

$$(1/2)D_1/(c + v)$$

and by that time the lead rocket's clock will have advanced to the reading,

$$t_0 + (1/2)\sqrt{1 - v^2/c^2}\, D_1/(c - v) + (1/2)\sqrt{1 - v^2/c^2}\, D_1/(c + v)$$

$$= t_0 + (1/2)\sqrt{1 - v^2/c^2}\, D_1\, [1/(c - v) + 1/(c + v)]$$

$$= t_0 + (D_1/c)/\sqrt{1 - v^2/c^2} = t_0 + D/c$$

Regarding 'trailing' rocket, the calculation just interchanges $c - v$ and $c + v$, so the result is the same: the two clocks have the same reading when the start signal is received. (This is just what synchronization implies.) However, the point is that in this case the start signal actually arrives at the 'lead' rocket before it arrives at the 'trailer': the aether time difference is

$$\Delta T = (1/2)D_1/(c - v) - (1/2)D_1/(c + v) = (D_1 v/c^2)/[1 - v^2/c^2]$$

$$= (Dv/c^2)/\sqrt{1 - v^2/c^2}$$

For that entire interval lead's clock will be advancing at the full rate since it is at rest in the aether, so when the start signal finally arrives at the trailer, lead's clock will read

$$t_0 + D/c + (Dv/c^2)/\sqrt{1 - v^2/c^2} \qquad \text{while A's reads} \qquad t_0 + D/c.$$

Thus again, lead's clock will be ahead of trailer's by the interval

$$\Delta T_2' = (Dv/c^2)/\sqrt{1 - v^2/c^2}$$

Regarding the separation distance, we merely note that during the interval ΔT, the separation increases to

$$D\sqrt{1 - v^2/c^2} + v\Delta T = D\sqrt{1 - v^2/c^2} + (Dv^2/c^2)/\sqrt{1 - v^2/c^2} = D/\sqrt{1 - v^2/c^2}$$

Again, since the rocketeers are at rest in the aether, $D_2' = D/\sqrt{1 - v^2/c^2}$

It is interesting to note that the actual initial and final separation distance, that is, the separations as measured by the aether observers, are not the same for the two cases considered. For the first case , $D_1 = D$ and $D_2 = D$, whereas, in the second case,

$$D_1 = D\sqrt{1 - v^2/c^2} \quad \text{and} \quad D_2 = D/\sqrt{1 - v^2/c^2}$$

Other topics of interest relegated to appendices
I can well imagine that at this stage the reader has had quite enough of mathematical manipulation. Acknowledging this, I have relegated some other interesting problems of special relativity to appendices.

Reprise regarding the limitations of special relativity
Before leaving the topic of special relativity, I should like to emphasize the fact that I am **not** asserting that standard relativity is incorrect. It is certainly not incorrect: it is in fact the only reasonable choice of the alternatives.[20] It is wonderfully convenient, and even very beautiful. But I do contend that the applicability of the theory is limited to situations in which synchronization is not fundamentally involved.

---

[20] The Lorentz transformation is just a special case of a more general transformation developed in 1977 by Mansouri and Sexl. Their more general formulation correctly predicts all the known physics, and differs from the Lorentz transformation only in regard to the convention regarding the synchronization of clocks, which they leave open. cf. R Mansouri and R U Sexl (1977) *Gen. Rel. Grav.* Vol.8, No.7, 497-524 and Vol. 8, No.10, 809-814.

Part II Einstein's Theory of Gravity

Part II (a) Einstein's dream of a General Theory of Relativity undone by Newton's bucket

Einstein continued to believe that it might be possible to develop an extension of his relativity theory that would embrace not just inertial (unaccelerated) systems, but any and all systems of observers, regardless of the their motion, including accelerating and rotating systems.

It is my firm view that Newton long ago had quashed any such ambition with his famous 'bucket' thought experiment.   Think of a bucket of water resting on the laboratory floor.   If the surface of the water is flat, the laboratory is not rotating.   Otherwise, the surface will be concave to some degree, and the laboratory must be rotating. If so, put the bucket on a turntable. One may determine the direction of motion by rotating the turntable. If the concavity increases, then the original rotation was in the same sense as was applied to the turntable.   In this case, turning the turntable in the opposite sense will eventually flatten the surface, and the rate of rotation of the turntable relative to the laboratory will be equal and opposite to the rotation rate of the laboratory.

Rotation is absolute in the sense that it is always in relation to a single frame of reference; that of the 'fixed stars.' Doing physics on a rotating platform must account for that rotation, and the laws of mechanics cannot be the same as those pertaining to an inertial system.   One may, it is true, employ tensor theory so as to write the laws of mechanics in universal form, but this merely disguises the centrifugal and Coriolis forces as Christoffel symbols which emerge from covariant differentiation.

It would seem that Einstein fixed his attention on 'straight line' acceleration. In his popularization, *Relativity*, Einstein gives us a *Gedankenexperiment* arguing that a person in a 'chest' enjoying (or suffering) uniform acceleration with respect to the 'Galilean' space of special relativity, will be unable to distinguish his situation from that of being at rest but suspended in a gravitational field. For the moment, we grant that without quibble. (At a later stage we shall indeed quibble, proving that the Equivalence Principle actually fails in the relativistic regime.) Einstein then asserts that it should always be possible to describe some fictitious, *ad hoc* gravitational field to permit an observer subject to some arbitrary but real episode of acceleration to assert that he is at rest, and the effects experienced are due to this *Deus ex machina* gravitational field. Neither Einstein nor anyone else has ever explained how this idea could be used to advantage. He continues as follows:

> "We have thus good grounds for extending the principle of relativity to in-
> clude bodies of reference which are accelerated with respect to each other,

and as a result we have gained a powerful argument for a generalized postulate of relativity."

In my view, this is a *non sequitur*, and the term 'general relativity' I take to be a misnomer.

Not surprisingly, he offers no suggestion as to how to actually formulate such a 'generalized postulate of relativity.' Instead he turns immediately to inferences deduced on the basis of the equivalence of linear acceleration and a constant gravitational field, inferences that were presented in his important paper of 1911. This paper is of great interest, not just historically, but because of an understandable failure on Einstein's part to fully appreciate the consequences of his reasoning, and the long range consequences of this failure.

Einstein's 1911 paper[21]

This paper is essentially an improved version of a review article published in 1907.[22] It begins with a comparison of an accelerated frame in field free space with a stationary frame in a gravitational field. Einstein notes that if the acceleration of the first frame K' is equal to the acceleration of gravity in the second frame, K, then the situations are equivalent with respect to Newtonian mechanics. He then goes on to explore the consequences of assuming that the two frames are equivalent with respect to all phenomena. This is Einstein's revered **principle of equivalence**.

Two proofs that gravitational mass and inertial mass are equivalent
Einstein in his special relativity had established that if energy E is added to a system, the *inertial* mass will be increased by the amount E / c^2. In the first part of this paper he sets out to prove that the same relation holds regarding *gravitational* mass.

Two observers S_1 and S_2 are equipped with identical devices to measure energy. They agree on a nominal amount of energy E to be sent as radiation from S_2 to S_1 when S_2 is at a distance h above S_1.

In the system K', we imagine that both observers are accelerating but are momentarily at rest when the radiation is emitted by S_2. When it arrives at S_1, that observer will have attained the speed $v = g(h / c)$, and the radiation will be blue shifted by the factor

---

[21] Einstein A 1911 *Annalen der physik* **35** translated as 'On the influence of gravitation on the propagation of light' in *The Principle of Relativity* 1952 (Dover) pp.97 -108.

[22] Einstein A 1907 *Jahrbuch der Radioaktivitat und Elektronik,* **4**, 411

$$\sqrt{\frac{1+v/c}{1-v/c}} \approx 1+v/c = 1+gh/c^2$$

If the extended principle of equivalence holds, the same procedures in system K in a gravitational field will lead to the same result, so that the energy, E_1 measured by S_1 will be

$$E_1 = E\left(1+gh/c^2\right)$$

Einstein observes that conservation of energy demands that all of this energy was in fact present at S_2 before it was released, which in turn requires that we ascribe to the energy E a difference of potential energy equal to Egh/c^2 between S_1 and S_2.

By way of clarification, Einstein continues the *Gedanken* in K by having S_2 lower a mass M down to S_1, releasing energy Mgh. S_1 adds the energy E (as measured in S_1) to M and then lifts the resulting object (of mass M' to be determined) back up to S_2.
This restores the original situation except that energy in the amount $E(gh/c^2)$ has been transferred from S_2 to S_1. Invoking energy conservation, one has

$$M'gh - Mgh = E(gh/c^2), \quad \Rightarrow \quad M' - M = E/c^2$$

This shows that the energy contributed to the gravitational mass, which was what Einstein set out to prove.

Note that there is a degree of circularity to Einstein's argument. He tacitly assumes that the energy E (measured by S_1) when lifted to S_2 will be increased by the amount $E(gh/c^2)$ so as to be equal to E when measured by S_2.

We shall never know how might Einstein have answered the question "what if S_1 just sends E (as measured by S_1) up to S_2 as free radiation. Will it enjoy the same increase in energy, as it did as captive energy in your *Gedanken*? " The answer, as can be inferred from Einstein's own work in this very paper, is *no*.

Having belabored this proof, Einstein now tells us that a much simpler *Gedanken* can demonstrate the equivalence of inertial and gravitational mass. We are asked to imagine a mass hung by a spring balance in each of the systems K and K'. If energy is added to the mass in K', there will be an increase in **inertial** mass, as was established in the special relativity program. By the principle of equivalence, the same elongation of the spring will be seen in K when energy is added. But in this case, that elongation must be attributed to an increase in **gravitational** mass, QED.

Einstein proves the red shift is caused by the slowing of clocks

Next Einstein addresses the issue that is of paramount importance for our discussion, namely, the gravitational shift of spectra lines. Clearly, the analysis already presented describes this phenomenon: S_1 observes a blue shift of the radiation from S_2:

$$f_1 = f_2 \sqrt{\frac{1 + v/c}{1 - v/c}} \approx f_2(1 + v/c) = f_2(1 + gh/c^2)$$

Einstein remarks

> "On superficial consideration [this] seems to assert an absurdity. If there is continuous transmission of light from S_2 to S_1, how can any other number of periods per second arrive at S_1 than is emitted in S_2?"

The answer, he assures us, is simple; yet he goes on at considerable length to come to the conclusion that the clocks used to measure frequency must run at different rates when located at different potential levels, even though they are of identical construction and remain synchronous when together at one location. *The crux of the argument is that in a time-independent situation, such as that under consideration, the number of wavecrests between S_2 and S_1 must remain constant – for every wavecrest that enters at S_2, one must leave at S_1.* The *true* frequency of light[23] is constant along the light ray and the different measurements of frequency made by S_1 and S_2 must be attributed to the fact that clocks are systematically slowed in a gravitational field. And the foregoing also shows that emission frequencies are slowed by exactly the same factor. Thus a particular spectral line will always be measured to have the same frequency provided that the source of line and the clock used to measure the frequency are at the same location or at the same gravitational potential.

An inference not made: Light moves with constant energy, unaffected by gravity

Strangely, Einstein does not make a point of emphasizing that free radiation moves with constant frequency in a time-independent field. I cannot understand why he did not emphasize this important point. It shows that there is a fundamental difference between what might be called *captive* radiation[24] on the one hand, and *free* radiation, on the other. This issue is central to the re-interpretation of Einstein's theory of gravitation to be presented, and will be dealt with in depth in due course.

---

[23] The thoughtful reader will ask how *true* frequencies can be defined. We may imagine timing signals broadcast from a remotely located master clock. Since the frequency of free radiation is constant, this signal can serve to regulate all clocks regardless of their location in the gravitational field.

[24] It will be demonstrated in Appendix V, 'Photon in a Superconducting Box" that captive radiation possesses an equivalent rest mass equal to E/c^2, and therefore does gain potential energy (and rest mass!) when raised in a gravitational field.

Note that this answers the question we imagined putting to Einstein – If S_1 had sent the energy E as free radiation, it would have been measured by S_2 to be equal to $E_2 = E(1 - gh/c^2)$ rather than E.

Clock slowing and the locally measured speed of light
This brings us to a crucial choice made, perhaps unconsciously, by Einstein. It was clear to him that having clocks run at various rates in a gravitational field would put in jeopardy the fundamental postulate of special relativity, namely, the constancy of the locally measured speed of light.

Einstein's unhappy choice: Light speed slows
Einstein assumes that the constancy of the speed of light must be maintained when measured using clocks of identical construction.[25] Such clocks are slowed by the factor $(1 + \Phi/c^2)$, where Φ is the gravitational potential, negative if infinity is taken as reference. Without further discussion Einstein asserts that the speed of light must be slowed by the same factor:

$$c = c_\infty (1 + \Phi/c^2)$$

Regarding this Einstein observes, almost apologetically,

> "The principle of the constancy of the velocity of light holds good according to this theory in a different form from that which usually underlies the ordinary theory of relativity."

In the final section Einstein employs this result to predict and calculate the deflection of starlight by the field of the Sun.

An alternate solution: Measuring rods elongate
But there was an alternative solution to the problem of maintaining the constancy of the measured speed of light. Speed is measured by noting the time required to traverse a course of known length. If the measuring rods used to determine the length of the course were to be elongated by a factor equal to the reciprocal of the factor by which clocks are slowed, then the measured speed of light would be constant, provided that the true speed of light were in fact constant. As we shall see, this turns out to be the case.

---

[25] Einstein asserted that in the system K' the measured speed of light would be constant when measured using clocks of identical construction, and therefore, by the principle of equivalence, the same had to be true in the system K. Einstein envisioned a very simplistic model for the system K', but in a more realistic model, described in Appendix VIII titled 'Einstein's Elevator', it will be shown that his assumption regarding K' was incorrect.

36

Bohr's 1913 quantum theory gives paradigms for time and distance intervals: $P_R = h^3 / m_e e^4$ and $a_0 = h^2 / m_e e^2$

In 1911, there was no fundamental theory to describe the characteristic frequencies and dimensions of atoms, or, what is directly related, that of clocks and measuring rods. But in 1913, Bohr's elementary quantum treatment of the hydrogen atom was published.[26] As will be discussed at length in the next section, this theory makes possible a systematic description of how the rates of clocks and the length of measuring rods are dependent upon just three fundamental 'constants,' namely, Planck's constant, h;

the mass of the electron, m_e; and the fundamental electric charge e.

As this is not a detective novel, it seems permissible to let the cat out of the bag: Bohr's theory showed that the characteristic frequencies emitted by the hydrogen atom are directly proportional to the electron mass, and that the diameter of the hydrogen atom is inversely proportional to the electron mass.

Understanding why rest masses are reduced in a gravitational field only requires a simple insight: *when an object is raised in a gravitational field, the gravitational potential energy increase is real, and exists as an increase, usually tiny, in the rest mass of the object.*

A heuristic argument for gravitational rest mass reduction

The arguments to be presented are based on quantum mechanics and may seem somewhat abstruse and removed from ordinary experience, but actually the reduction of rest masses is easily understood. Lifting a heavy object requires the expenditure of energy. But the energy invested is recoverable – it exists, we have been told, as 'potential energy.' Where is this energy? Presently, physicists answer, vaguely, "It's in the gravitational field, manifest as a decrease in the *negative energy* of the field." Say what?! If this strikes you as reminiscent of medieval theology, you're right. In fact, the potential energy is in the object itself, existing as an increase, usually very tiny, in the rest mass of the object, in accord with the famous equation, $E = mc^2$. Thus if work, ΔE, is done on a body, the mass of that body will increase by the amount, $\Delta m = \Delta E / c^2$. Conversely, lowering an object (and absorbing the released potential energy) will result in a decrease in the rest mass of the object.

A Self-consistent derivation of rest mass reduction is presented in Appendix XII.

---

[26] Niels Bohr, *Philosophical Magazine* **6**, 26 July 1913, pp. 1-25

The implications of rest mass reduction

The fundamental insight offered here is that *rest masses are reduced in a gravitational field.* This single phenomenon accounts for the red shift (aka - clock slowing) and also explains how the speed of light is a measured constant in spite of the slowing of clock rates. Most importantly, it reveals that the accepted spatial geometry, which is defined in terms of measurements made with actual (elongated) measuring rods, does not correctly describe the true geometry. In the case of a black hole, the true geometry is radically different from the currently accepted geometry.

As you may know, the prediction made by Einstein in this paper for the deflection of light gave only half the value later calculated on the basis of his general theory. The reason is that, at this stage, Einstein had not conceived of the possibility that non-Euclidean geometry would be involved. As we shall show, it turns out that the whole deflection must be attributed to the curvature of the true geometry, which curvature is in a sense twice that of the geometry of the standard interpretation.

Einstein's struggle to develop his theory of gravity

This history is most complex and difficult to describe. The rudimentary description below is borrowed mainly from the very thorough account of Abraham Pais.[27]
In 1912 Einstein published two papers on gravity. Having convinced himself of how the velocity of light must be related to gravitational potential, he attempted to develop a theory of an external static "c-field," c being the speed of light. The interesting thing about this work is that Einstein was able to show that an equation for the motion of a test particle could be derived from a variational principle, namely,

$$\delta \int ds = 0 \quad \text{where} \quad ds^2 = c^2 dt^2 - dx^2 - dy^2 - dz^2$$

This prefigured the final theory in which the differential invariant is generalized to Riemannian space time:

$$\delta \int ds = 0 \quad \text{where} \quad ds^2 = g_{jk} dx^j dx^k$$

In the same year Einstein concluded that the assumption of flat space was untenable. His reasoning followed from his confidence in the possibility of an actual theory of general relativity. Such a theory when applied to a uniformly rotating system would have to deal with non-Euclidean geometry since the FitzGerald contraction would imply that the ratio of the circumference to the radius would be greater than 2π.

---

[27] Abraham Pais, *'Subtle is the Lord...' The Science and the Life of Albert Einstein.* Oxford University Press (1982)

38

Marcel Grossmann and Einstein

Einstein only became aware of Riemannian geometry or the work of Ricci and Levi-Civita when he posed to his friend Marcel Grossman the problem of finding

> "…generally covariant tensors whose components depend only on derivatives of the coefficients [g_{jk}] of the quadratic fundamental invariant [$g_{jk}dx^j dx^k$]."

Einstein considered the g_{jk} to be dynamic fields which somehow described gravitation, and Grossman assured him that what was needed was Riemann's geometry. In co-authored paper [28] Grossmann wrote

> "The problem of the formulation of the differential equations of a gravitational field draws attention to the differential invariants… and… covariants of $ds^2 = g_{jk}dx^j dx^k$."

Grossmann immediately recommended the Riemann-Christoffel Tensor, R^i_{jkl} and its contraction, $R_{jk} = R^l_{jkl}$, the Ricci tensor. This put the program on the right track, but there remained misunderstandings to be overcome. One such was Einstein's mistaken idea that the metric coefficients g_{jk} had to be uniquely determined, whereas in fact, they are determined only up to an arbitrary coordinate transformation. Finally in 1915 Einstein derived the equation

$$R_{jk} - \frac{1}{2}g_{jk} R = \frac{8\pi G}{c^4} T_{jk}$$

Here T_{jk} is the matter tensor. The tensor on the left, now called the Einstein tensor, has the property (by the Bianchi identities)

$$(R^{jk} - \frac{1}{2}g^{jk} R)_{;k} = 0$$

so that the entire equation is compatible with the conservation laws for matter

$$T^{jk}_{\ ;k} = 0$$

---

[28] A Einstein and M Grossmann, Z. Math. Physik. **62**, 225 (1913)

(The semi-colon subscript represents the covariant derivative operator.)

Fortunately, a deep understanding of the mathematics involved in Einstein's theory of gravity is not required for the purposes of the present writing. We shall only be dealing with classical solutions – that of Karl Schwarzschild describing a time-independent and spherically symmetric field, and that due to Roy Kerr, which describes the time-independent field external to an axi-symmetric rotating body.

Incidentally, it should be noted that in Einstein's theory of gravity, the coordinates are chosen as a matter of convenience, and do not generally represent physical measurements with real clocks and measuring rods, as was the case in the special theory of relativity. Some theorists, believing that Einstein's 'general relativity' program actually succeeded, have misunderstood this situation, and assert that the arbitrariness of coordinate choices indicates that the theory embodies a transformation group that is much broader than the Lorentz group of the special theory. This is not the case.

40

Part II. (b) The Variable Rest Mass Interpretation of Einstein's Theory of Gravitation

Introduction

Today many relativists are of the opinion that any interpretation of the theory is superfluous, being merely a matter of taste. If one seeks understanding, they assert, one need only consider the mathematics. Despite such protestations, nearly all of these individuals actually subscribe, indeed rather dogmatically, to a certain interpretation of Einstein's theory. In this interpretation, the usual proper spatial metric, based upon proper measurements, is thought to correctly represent the geometry of space near a gravitating body. Also, the speed of light is commonly assumed to be reduced in a gravitational field[29]. But, as will be demonstrated, interpretations are not completely arbitrary, and, in fact, the presently accepted conventional interpretation is incorrect.

The Strong Equivalence Principle

The equivalence principle, improperly understood, has led some physicists to believe that the instruments used to make local measurements are unaffected by a gravitational field. This is incorrect. The *strong equivalence principle* (SEP) is a fundamental assumption of Einstein's theory. As formulated by R. H. Dicke[30], it asserts that

> "In a freely falling, non-rotating laboratory, the local laws of physics take on some standard form, including a standard numerical content, independent of the position of the laboratory in space and time." [31]

Fundamental 'constants' can vary in concert so as to satisfy the SEP

Thoughtfully considered, this statement *does* allow that the instruments used for the measurement of mass, length and time (and perhaps fundamental 'constants') may be dependent upon position in a gravitational field, *provided that they vary together in such a manner as to satisfy the SEP.*

---

[29] As discussed above, Einstein originated this concept in his 1911 paper, and he continued to accept it: cf. *Relativity*, 1st ed., 1916; 15th ed., 1952, p 76.

[30] Dicke R H *The Theoretical Significance of Experimental Relativity* (Gordon and Breach, New York, (1965) p.4

[31] This principle is really the underlying assumption regarding the whole enterprise of science. It is only vaguely related to Einstein's original Equivalence Principle, which only asserted that acceleration of magnitude, a, in Galilean space is equivalent in its local effect to being stationary in a gravitational field with acceleration of gravity, $g = a$.

The gravitational red shift, properly understood as being the result of a frequency reduction at the source, proves that, in fact, the standards for the measurement of time <u>do</u> vary with position in a gravitational field.[32] An immediate conclusion is that the other 'standards,' for the measurement of length or mass, or possibly, the true values of fundamental 'constants,' must also depend upon position in the field in some manner so as to satisfy the SEP.

The constancy of the speed of light, *as measured locally using local instruments*, is also fundamental to Einstein's theory. (With this postulate, Einstein extended Galilean relativity to include optics. Of course, it is implied by the SEP.) As has been remarked, Einstein realized that the slowing of clocks required that some other 'standard' had to vary in order to keep the locally measured speed of light constant. Einstein 'saved the postulate' by assuming that the *true* speed of light must be reduced by the same factor by which clock rates are reduced. This idea has persisted and is one of the features of the conventional interpretation. Thus in the present era, theorists find that the Shapiro effect (the delay of electromagnetic signals passing close to a gravitating body such as the Sun) cannot be completely accounted for in terms of the path length *as computed according to the usual proper metric*, which assumes that measuring rods are unaffected by gravity. These theorists conclude that the speed of light must be reduced in a gravitational field so as to bring theory in line with observation.[33]

Accordingly, we shall systematically investigate the manner in which these 'standards' can vary so as to satisfy the single known fact, the red shift, as well as the other basic requirement of the theory – the SEP, including the constancy of the locally-measured speed of light.

---

[32] Einstein finally clarified his view stating, "…we can regard an atom which is emitting spectral lines as a clock, so that the following statement will hold: *An atom absorbs or emits light of a frequency which is dependent on the potential of the gravitational field in which it is situated.*" [Emphasis Einstein's] A Einstein, *Relativity*, (Crown Publishers, New York, 1952), 15th ed., Appendix III(c), pp.130-131 in hardcover, p. 149 in paper.

[33] Schwinger J (1986) *Einstein's Legacy* (Scientific American Books, Inc) New York pp.204-06

Interpretations, correct and otherwise

In what follows, the **true** value of 'constants' will be indicated by appending an asterisk to the usual symbol. **Proper** values, which are numerical constants according to the SEP, will be written plain. To avoid unnecessary abstraction, attention will be confined to the Schwarzschild field, in which the usual (proper) metric is written

$$ds^2 = f^2 c^2 dt^2 - f^{-2} dr^2 - r^2 (d\theta^2 + \sin^2\theta \, d\phi^2)$$

Here r = the 'radius' = (proper circumference)/2π, and
$f = \sqrt{1 - r_S / r}$, in which $r_S = 2GM / c^2$, is the
'Schwarzschild radius.'

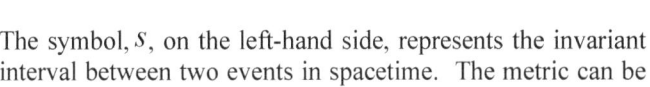

The symbol, s, on the left-hand side, represents the invariant interval between two events in spacetime. The metric can be thought of as a differential equation for s : for events separated by finite differences in the coordinate variables, one integrates to find s . For a clock at rest or moving between two events, s/c represents the **proper time** recorded by that clock. The proper time is distinct from the coordinate time, t, which is called the **world time**. For events occurring 'simultaneously,' i.e., at the same value of the world time, t, the invariant interval, $|s|$, represents the **proper distance** between the two points in space. For two events connected by a light signal, $s = 0$. Clearly, the metric shows that non-moving clocks are slowed by the factor, $f = \sqrt{1 - r_S / r}$, as compared with world time.

In the conventional interpretation, it has been assumed that $c^* / c = f$ in order that the locally measured speed of light remain constant. In general, for some quantity, x, we shall indicate its dependence upon the field by writing

$$x^* / x = f^{(x)}$$

in which the exponent, (x), may or may not be an integer. In the early days of relativity it was not clear how one could deal systematically with clock rates and the length of measuring rods. Since the advent of quantum mechanics, the way to proceed is obvious.

According to the SEP, clock rates and the frequency of spectral lines must vary together, and the length of a measuring rod must vary in lockstep with the diameter of atoms. Thus we may take as a standard for time what might be called the Rydberg period,

$$P_R = h^3 / m_e e^4$$

and for a length standard, the Bohr radius,

$$a_0 = h^2 / m_e e^2 .$$

The one experimental fact is the slowing of clock rates, which implies

$$P_R \sim f^{-1} \quad \Rightarrow \quad 3(h) - (m) - 2(e^2) = -1$$

The second constraint is that the locally measured speed of light, c_{meas}, must be a constant. Clearly, a slowing of clock rates would produce an increase in c_{meas}, while an elongation of measuring rods would reduce c_{meas}. And, of course, c_{meas} is directly proportional to the true value, $c*$. Thus we can write

$$c_{meas} = c*(P_R*/P_R)(a_0*/a_0)^{-1} = c*(P_R*/a_0*)(P_R/a_0)^{-1} = c*(h*/h)(e^2*/e^2)^{-1}$$

But the theory demands that $c_{meas} = c$, and so we have

$$(c*/c)(h*/h)(e^2*/e^2)^{-1} = f^0 \text{, i.e., } (c)+(h)=(e^2)$$

Interestingly, this says that the fine structure constant, $\alpha = e^2 / hc$, must be an actual constant if the locally measured speed of light is to be constant.

Of course, the fundamental unit of charge, e, must be an absolute invariant in order that charge be conserved. (Just think of a cat that rubs against an amber vase on the mantle and then jumps down, leaving an excess of electrons on the vase and net positive charge in its fur.) Thus there are only three variables – (c), (h) and (m) – to be determined, and just one more stipulation fixes the behavior of all three.

Einstein's choice versus variable rest masses
We'll begin with what might be called Einstein's choice (also the unconscious choice in the conventional interpretation), namely, that the lengths of measuring rods are unaffected in a gravitational field, whence

$$a_0 \sim f^0 \quad \Rightarrow \quad 2(h) = (m)$$

This determines everything, and yields

$$(a_0) = 0 \quad (c) = 1 \quad (h) = -1 \quad (m) = -2$$

One may well imagine that those subscribing to the conventional interpretation will be more than a little surprised to learn the full implications of their assumptions, namely, that $(a_0) = 0$ and $(c) = 1$, which they have casually accepted. That interpretation also tacitly assumes that $(h) = 0$, and $(m) = 0$, but this is clearly impossible, showing that the conventional interpretation is incorrect.

Next we consider the author's choice, which is supported by a number of considerations, including the conservation of energy, and, importantly, the capacity to explain the Abramowicz effect,[34] the reversal of the direction of centrifugal force inside the realm of photon orbits at $3/2\, r_S$.

The Abramowicz Effect and Optical Geometry

Permit a brief digression to explain the connection of the work of Abramowicz et al. with the theory here presented. Marek Abramowicz, Brandon Carter and Jean-Pierre Lasota proved that a body will not experience velocity-dependent forces (centrifugal and Coriolis forces) if it moves along a path that a light ray might follow. By analogy to Newtonian mechanics, in which such forces vanish for bodies moving in straight lines, the authors defined a new geometry for general relativity by identifying light  rays as the geodesics (the nearest thing to a straight line in non-Euclidean geometry) of the new geometry. This new geometry they named, appropriately, 'Optical Geometry' [35]. It is identical to the geometry presented in this book.

The variable rest mass interpretation simply asserts that

$$(m) = 1 \quad \Rightarrow \quad (a_0) = -1 \quad (h) = 0 \quad (c) = 0$$

It is interesting to see the implications of demanding that rest masses be invariant:

$$(m) = 0 \quad \Rightarrow \quad (a_0) = -2/3 \quad (h) = -1/3 \quad (c) = 1/3$$

[34] Abramowicz M A and Lasota J-P (1974) "A note of a paradoxical property of the Schwarzschild solution" *Acta Physica Polonica* B5, 327

Abramowicz M A and Lasota J-P (1986) "On traveling round without feeling it and uncurving curves" *Am. J. Phys.* **54** (10) 936-38

Abramowicz M A (1992) "Relativity of inwards and outwards: an example" *Mon. Not. Astr. Soc.* 710-18,

[35] Abramowicz M A, Carter B and Lasota J-P (1988) *Gen. Rel. Grav.* **20** 1173-83

Energy conservation picks the winner

Consideration of energy conservation provides the means of choosing the correct interpretation. The Schwarzschild field is a time-independent field, and in such a field, energy must be conserved. There is an exact integral of motion in this field, namely,[36]

$$mc^2 f / \sqrt{1 - v^2/c^2} = \text{constant}$$

The correct interpretation is that in which $m*c*^2 = mc^2 f$, so that the integral of motion expresses energy conservation in terms of the standard expression for energy, namely,

$$m*c*^2 / \sqrt{1 - v*^2/c*^2} = \text{constant}$$

In Einstein's interpretation, $m*c*^2 = (mf^{-2})(cf)^2 = mc^2$, which fails the test. The same is true for the $(m) = 0$ option, where $m*c*^2 = (mf^0)(cf^{1/3})^2 = mc^2 f^{2/3}$. For the scheme invoking rest mass reduction, $m*c*^2 = (mf)(cf^0)^2 = mc^2 f$, as required. Finally, there may be concern regarding the relativity factor, v/c, in the first equation above, in which the velocity, v, is the proper velocity. But according to the rest mass reduction interpretation, measured time and length intervals are underestimated by the same factor, f, and hence the proper measure of any velocity is correct, i.e., $v = v*$. Of course, $c* = c$ as well.

The energy equation immediately above clearly shows that the gravitational potential energy of a body of proper rest mass, m, exists in the body itself, as a increase or decrease in the true rest mass, $m*$, of the body.

Actually, the red shift provides a simpler and equally effective test. As already noted, the red shift results from a reduction of frequency of the source, and this means that the frequency 'in flight,' so to speak, is constant. Call this constant frequency $v*$. Energy conservation then simply demands that $h*v* = \text{constant}$, and this in turn requires that $h*$ be constant, a condition satisfied only for the interpretation in which $(m) = 1$, which will subsequently be referred to as the variable rest mass (VRM) interpretation.

In fact, the simple assumption, $(h) = 0$, says it all. Equation (1) and charge invariance requires $(m) = 1$. Then from the definition of the Bohr radius, $(a_0) = -1$.

---

[36] Landau L D and Lifshitz E M (1962) *The Classical Theory of Fields, rev 2nd ed.* Pergamon Press, p.292

The New Metric and the True Geometry

According to the VRM interpretation, proper time intervals and distance intervals are underestimated by the same factor, $f = \sqrt{1 - r_S / r}$. Correcting for these deficiencies is thus very simple: one only need multiply the proper metric by the factor f^{-2}. The metric in the correct interpretation is thus related to the proper metric by a simple conformal factor. Writing $s*$ for the interval in the VRM metric, we may write

$$ds*^2 = f^{-2}ds^2 = c^2 dt^2 - f^{-4}dr^2 - f^{-2}r^2(d\theta^2 + \sin^2\theta \, d\phi^2)$$

At this point it is appropriate to introduce some notation and to discuss at some length the character of the new geometry.

First of all, one may introduce a system of space-time measurement that is consonant with the VRM interpretation. Early on we emphasized the importance of the fact that freely moving electromagnetic waves move with constant frequency. This is the key to the implementation of a system of space-time measurement that is unaffected by the gravitational field. One simply measures time using the signals from a single remote clock. (This is identical to the 'world time' t.) Distance measurements are then made (by local or remotely located observers) using electromagnetic echo ranging (radar) techniques, calculated using the time as measured, not by a local clock, but by the same remote clock.

Since the new system of measurement employs a remote clock and makes use of a remote sensing technique for distance measurements, the name **telemetric** is sug- gested. Alternately, the new system may be deserving of the label '**true**.' Even '**transcendent**' seems not inappropriate. Allowing for any of these, the new system will sometimes be referred to as the 'T system,' and the * metric above may be called the 'T metric.'

Perhaps the most striking thing about the T metric is that the force of gravity has vanished. The integral of motion above may be written more generally for any static field as

$$mc^2 \sqrt{g_{00}} / \sqrt{1 - v^2/c^2} = \text{constant}$$

But in the T metric, $g_{00} = 1$, and hence the speed v is a constant. The geometry is, however, quite non-Euclidean. Quite generally, whenever g_{00} is independent of position, bodies will move along spatial geodesics at constant speed. This puts the lie to the ubiquitous and absurd depiction of gravity as resulting from the curvature of space – the 'bowling ball on the rubber sheet' mal-analogy. (see Appendix XIII, section 3)

The Stenosphere

In the T metric, the area of a centered sphere is equal to

$$4\pi r^2 f^{-2} = 4\pi r^2 (1 - r_S / r)^{-1},$$

Differentiating with respect to r, one has

$$d / dr [r^2 (1 - r_S / r)^{-1}] = (2r - 3r_S)(1 - r_S / r)^{-2}$$

Thus the area of the sphere is not a monotone function of r: it has a minimum at $r = 3/2 \ r_S$, the locus of the photon orbits. Furthermore, for $r < 3/2 \ r_S$, the area of the sphere increases without limit as r approaches r_S. The sphere of minimum area, the **stenosphere**, is the throat of a wormhole-like structure connecting our familiar universe with another infinite three-space, which may be called '**innerspace**.' The VRM interpretation describes a geometry that is radically different from that currently accepted by most physicists.

In particular, note that inside the stenosphere, the surface of a centered sphere, viewed from the 'outside,' will be concave rather than convex. Thus, regarding the Abramowicz effect, the correct geometry shows that the direction of centrifugal force obeys the usual pattern inside the stenosphere: the force is directed from the concave side to the convex side of the circle on which a body is constrained to move.

Note that minimizing the distance between two points in space using the T spatial metric

$$dl *^2 = f^{-4} dr^2 + f^{-2} r^2 [d\theta^2 + \sin^2 \theta \, d\phi^2]$$

is identical to setting $ds = 0$ in the usual space-time metric

$$ds^2 = f^2 c^2 dt^2 - f^{-2} dr^2 - r^2 [d\theta^2 + \sin^2 \theta \, d\phi^2]$$

and minimizing the world time for light to traverse the interval between the same two points. Thus light rays move along the spatial geodesics of the T geometry. In contrast, proper geodesics are defined as the shortest paths as measured by local observers using local measuring rods; these 'shortest' paths in proper measure 'cheat' by deviating inwardly so as to take advantage of the length dilation effect, which results in a smaller value for the distance measured. This explains the fact that in the usual interpretation, based as it is on the proper metric, light rays do not follow geodesics.

Embedding diagram(s) for the Schwarzschild field

In order to clarify the nature of the geometry of the Schwarzschild field, it will be helpful to introduce what is called an embedding diagram. In this diagram, the section $\theta = \pi/2$ will be shown (all central sections are identical). In the absence of the central mass that produces the field, such a section would be a simple Euclidean plane, but with the mass present, the surface must be permitted to bulge up or down – down is the usual choice – so as to correctly depict the relation between small distances, $dR*$ and $r*d\phi$, in the radial and circumferential directions, respectively.

Because of the spherical symmetry, the surface is a figure of revolution of which only the right half of a vertical section will be shown in fig.1.

In this diagram, $r*$ ($1/2\pi$ times the telemetric circumference of circles) is the abscissa; the parameter, h, which has no direct meaning, is the ordinate; $R*$ is the telemetric distance along the surface in the radial direction; and the coordinate, r, now serves merely as a convenient parameter.

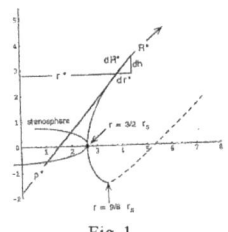

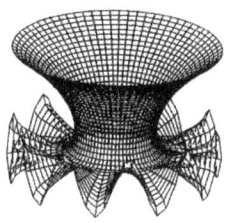

Fig. 1
Telemetric embedding diagram for the Schwarzschild field: right half of cross-section.

Fig. 2
Telemetric embedding diagram for the Schwarzschild field.

It is important to understand that points off the diagram surface are not part of the three-space of the Schwarzschild field, and that, regarding the curvature evident in the diagram, only that component lying in the tangent cone is sensible to observers in the three-space.

The differential equation for the surface in the embedding diagram is developed from the relation $dh^2 = dR*^2 - dr*^2$, as indicated in figure 1, along with the relations $r* = f^{-1}r$ and $dR* = f^{-2}dr$. The result expressed in terms of the parameter r is

$$dh/dr = f^{-3}\sqrt{\frac{2r_S}{r}\left[1 - \frac{9}{8}\frac{r_S}{r}\right]}$$

It will be noted that when $r < (9/8)r_S$, the expression for dh/dr becomes imaginary. This is a consequence of the fact that in the infinite region $(9/8)r_S > r > r_S$, $r*$ increases faster than $R*$. In this region, the embedding diagram should actually have the appearance of a floppy hat, with a circumferential excess of material, as shown in figure 2.

In figure 1, for the sake of clarity, the imaginary values for h are accepted, and are indicated by the dashed part of the curve in the diagram.

The sensible radius of curvature, $\rho*$, is the distance from the point in question along the tangent of the surface to the central axis. On the stenosphere, $\rho*$ becomes infinite, indicating that this surface is indeed flat. This is consistent with the well-known fact that (unstable) photon orbits occur at this location, $r = (3/2)r_S$.

David Hilbert

50

Note also that a circle when viewed from what is conventionally denoted as the 'outside' will be judged convex if it is outside the stenosphere, but will appear to be concave if it is inside the stenosphere.

The figure below compares the embedding diagrams for the telemetric and proper geometries. As is evident, the former is much more curved.

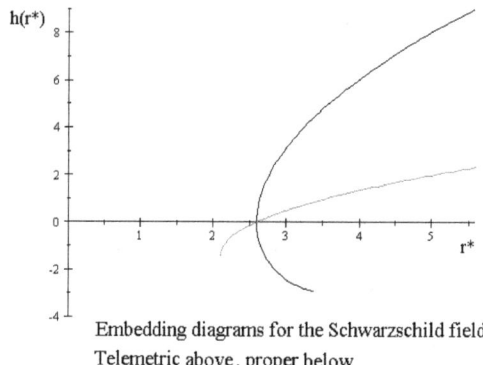

Embedding diagrams for the Schwarzschild field:
Telemetric above, proper below

It is important to stress that these diagrams depict information about the geometry of the space: they do not represent gravitational potentials.

Exploring the geometry of a black hole

The swirling-cone depiction of a black hole, which has become a cliché, is hopelessly misleading. Understanding requires that one abandon simple pictures depicting the black hole as an object in ordinary three-dimensional space: one must, I believe, *experience* the geometry.

Non-Euclidean geometry is not easy to grasp, but useful insights can be gained by what might be called a ***Gedankenerforschung*** (thought exploration). Imagine that you and nineteen compatriots are piloting small but very powerful rocket ships in an exploration of a black hole.[37] Using their rocket thrusters, the ships can hover as desired against the gravitational force of the black hole. The ships take up a symmetrical formation surrounding the black hole, with one ship stationed on each vertex of a dodecahedron.

As the ships move closer, the area of the sphere on which the ships are located decreases, but rather less than would be expected in flat Euclidean space. It seems there is an ***excess of space*** near the black hole. The discrepancy grows worse as the hole is closely approached, and at a certain distance (1.5 times the Schwarzschild radius, r_S) the area of the sphere on which the ships are located remains constant as the ships move slightly closer! Moving even closer, the area of the sphere is seen to actually increase! Not only that: the surface of the sphere, looking inward, is now seen to be ***concave*** rather than convex – the whole cadre is now inside the sphere! Also, the observers will be surprised to see that the ships which had been visible ahead are now seen behind. The sphere on which the observers positioned themselves has been everted![38] The sphere of minimum area may be called the stenosphere (Greek *stenos* = narrow).

Looking ahead, there seems to be no barrier to further inward progress: the ships have entered a new infinite space, connected to our familiar universe by a magical doorway, the stenosphere. The event horizon at r_S is not to be seen, for the very good reason that it is infinitely distant in the newly discovered universe.

Now the expedition retraces its inward journey, moving back out in order to perform some experiments. At some distance outside the stenosphere, one of the ships first hovers, noting the thrust required to maintain its distance. Then, using side thrust, it begins to move around the black hole. As the tangential velocity is increased, it requires less radial thrust in order to maintain the radial distance. Finally, moving

---

[37] We shall imagine that neither the explorers nor their rocket ships are subject to the effects of rest mass reduction.

[38] Smale (1958) proved that it is mathematically possible to turn a sphere inside-out without introducing a sharp crease at any point. In 1961, Arnold Shapiro devised an explicit eversion but did not publicize it. Phillips (1966) heard of the result and, in trying to reproduce it, actually devised an independent method of his own. Yet another eversion was devised by Morin, which became the basis for the movie by Max (1977). Morin's eversion also produced explicit algebraic equations describing the process. The original method of Shapiro was subsequently published by Francis and Morin (1979). [Credit to Wikipedia.]

sufficiently fast around the black hole, thrust in the radial direction is no longer required to maintain distance: the ship is in a natural orbit. Increasing the tangential velocity further now requires reversing the thrust to produce an inward force in order to maintain distance: centrifugal force now exceeds the force of gravity. This is just what is one expects.

The reversal of centrifugal force: the Abramowicz Effect

The same experiment is now carried out at distance $1.5\,r_S$ – right on the stenosphere. Here it is found that the tangential velocity has no influence on the radial thrust required to maintain distance: at this location centrifugal force has vanished! And when the experiment is repeated inside the stenosphere, it is found that any increase in the tangential velocity requires an *increase* in radial thrust, indicating that the direction of centrifugal force has reversed. This seems counter-intuitive, but on reflection, we realized that the centrifugal force is acting as it always does: the force is directed from the concave side to convex side of the circle on which the ship moves.

This vanishing and reversal of centrifugal force, the Abramowicz Effect, already mentioned above, is totally inexplicable within the context of the proper geometry, but is obvious in terms of the correct geometry. Likewise, the photon orbits that occur on the stenosphere are understandable considering that this sphere is a surface of zero intrinsic curvature, and photons, as we have seen, are not subject to the force of gravity.

According to the present standard interpretation, it is accepted that a real space ship, subject to rest mass reduction, might plunge into the black hole, penetrating the event horizon in a finite (in fact, a rather short) period of time – *as measured by an on-board clock*. Such penetration is held to be possible in spite of the fact that no stationary observer outside the event horizon will ever witness this happening: the time required for the ship to reach the event horizon, even as measured by slow-running local clocks, is infinite. This should be quite puzzling to local observers, since the proper distance to be traversed is finite, and the speed of the ship rapidly approaches the speed of light. In fact, even a photon cannot reach the event horizon in any finite time. [39]

But in fact, the geometry inferred from proper measurements is untenable. The supposedly finite proper distance to the event horizon is an illusion resulting from the fact that real measuring rods would become infinitely long as $r \to r_S$. Also, the finite time recorded by the on-board clock is likewise illusory, since the rate of such a clock will swiftly approach zero.

---

[39] For a photon in radial infall from r_0, $\quad ds = 0 \quad \Rightarrow \quad dt = -c^{-1} f^{-2} dr$

$$\Rightarrow \quad t = c^{-1}[(r_0 - r) + r_S \ln[(r_0 - r_S)/(r - r_S)] \to \infty \quad \text{as} \quad r \to r_S$$

In the telemetric picture, the 'black hole' is not necessarily black, and the 'hole' is a sort of 'worm hole' whose 'throat' is the stenosphere located at $1.5\ r_S$. It connects our familiar universe with another infinite three-space, which may be called 'innerspace.' A collapsing star does not suffer an infinite compression, producing a singularity hidden behind an event horizon; it produces a sort of rupture in space, forming a connection to an 'innerspace' into which it falls.

An intriguing question arises regarding black holes: namely, whether two or more black holes share a common innerspace, or whether each such is a separate, infinite three-space. If one innerspace serves for all, then the possibility emerges, however impractical it may be, of travel from one black hole to another via an innerspace route.

Regarding Quantum Gravity

The question as to the nature of the supposed graviton is regarded as settled: the graviton is taken to be the quantum of a massless spin-two field. The arguments against a spin-zero graviton are all based upon the seeming fact that photons couple with the gravitational field, which would be impossible in the case of a spin-zero graviton. (Spin zero particles couple only to the trace of the energy-momentum tensor of the field, which trace vanishes in the case of the electromagnetic field.) The supposed coupling is thought to be demonstrated by the experimentally observed deflection of light by massive objects such the Sun or, in the case of gravitational lensing, a whole galaxy, complete with its dark matter halo.

But we have seen that in the context of the correct telemetric (optical) geometry, the deflection of light rays is wholly due to the spatial curvature associated with a gravitational field. Light rays simply follow the spatial geodesics of the correct geometry. We have also seen that the energy of freely moving photons is unaffected by gravity. Thus the rejection of the spin zero graviton on the ground of the supposed coupling of light and gravity is a mistake.

More fundamentally, the universally accepted idea the mass and energy are in all ways equivalent leads to the idea that gravity couples with every form of mass-energy. But the proper understanding of the gravitational red shift proves that freely moving photons do so with constant energy, and so are unaffected by gravity. Gravity does not couple with massless particles.

Beside the obvious benefit of being described by a much simpler theory, the spin-zero graviton, being massless, will not couple with itself, eliminating the dreaded non-linearity of the proposed spin-two theory. The spin-zero graviton will also solve the vexing problem of the stubborn infinity that accompanies attempts to calculate the zero-point energy of the vacuum when gravity is combined with other quantized fields.

Digressions of particular interest

It seems reasonable at this point to treat two issues of considerable importance. The first concerns the nature of the gravitational 'constant', G. It has long been recognized that if G is actually a constant, then Einstein's as well as Newton's theory of gravitation would fail to satisfy the demands of the Strong Equivalence Principle. We shall argue that G must be dependent upon the gravitational potential: $G*/G = f^{-2}$ in the case of the Schwarzschild field.

The second important issue concerns the question as to whether Einstein's original equivalence principle is actually correct. It will be shown that, ironically, it is correct only in the non-relativistic regime, and is quite incorrect with respect to the deflection of light, the prediction of which has been much celebrated. To spare those readers who do not relish math tedium, the results are summarized here. The critique is based upon calculating the radius of curvature of the particle or photon path as it traverses the elevator imagined in Einstein's famous *Gedanken*. Einstein's analysis predicts the radius of curvature to be

$R = v^2/g$ for particles of velocity $v = \beta c$, and for photons, ($v \to c$) as well.

In contrast, detailed analysis gives, in the case of the Schwarzschild gravitational field,

$$R = \frac{v^2}{g}\left(\frac{1}{1+2\beta^2}\right) \text{ for particles and } \quad R = \frac{1}{3}\frac{c^2}{g} \text{ for photons.}$$

The reason for the discrepancy lies in the fact that a real gravitational field induces spatial curvature, a phenomenon that Einstein did not anticipate when in 1907 he formulated the principle of equivalence.

Big G: Non-constant and split!

If we accept the Strong Equivalence Principle (SEP) in every context, then we expect that the result of the Cavendish-type experiment would be independent of the location in a gravitational field: all such experiments would give the same value for Big G, the so-called Newtonian constant of gravity. (Perhaps experiments of the Cavendish type might be carried out on the International Space Station to test the validity of the SEP.)

Adherence to the SEP will require that the equation for the force of gravity between two masses, separated by a small distance, will be invariant under the changes of the rate of clocks and the length of measuring rods that occur in a gravitational field.

$$m_1 \frac{d^2 r_1}{dt^2} = G \frac{m_1 m_2}{(r_1 + r_2)^2}$$

Recall our convention regarding the dependence of some quantity, x, upon the field by writing $x*/x = f^{(x)}$. Then adherence to the demand of the SEP requires

$$(G) = 3(l) - 2(t) - (m) = 3[-1] - 2[-1] - [1] = -2$$

Thus in the Schwarzschild field, $\qquad G*/G = f^{-2}$ where $f = \sqrt{1 - r_S/r}$

This is adequate as regards two masses close together as in a Cavendish experiment. But what if we consider two small asteroids orbiting at different distances, r_1 and r_2, in the gravitational field of a large star. What value of G are we to assign regarding their gravitational interaction?

One possibility is to 'split' G, assigning $\sqrt{G*(r_1)} = \sqrt{G}f^{-1}(r_1)$ to the body at r_1 and

$\sqrt{G*(r_2)} = \sqrt{G}f^{-1}(r_2)$ to the body at r_2. But notice that

$$\sqrt{G*(r)}\, m*(r) = \sqrt{G}f^{-1}(r)\, m f(r) = \sqrt{G}\, m \text{ , a constant for the body.}$$

Thus the (true) Newtonian force acting between the bodies reverts to the familiar form

$$F* = G m_1 m_2 / R*^2 (r_1, r_2)$$

Where $R*(r_1, r_2)$ is the true distance between the bodies.

It may be useful to assign a **constant** 'gravitational mass' to a body equal to

$$m_G = \sqrt{G}\, m .$$

Then the force between the masses would be simply

$$F* = m_{G1} m_{G2} / R*^2 (r_1, r_2)$$

But note that at any location, r, $F(r)*/F(r) = f^2(r)$: hence the **proper** forces measured at r_1 and r_2 are not equal: rather

$$F(r_1)/F(r_2) = [f(r_1)/f(r_2)]^{-2}$$

We shall revisit big G in our consideration of cosmology – namely as to whether gravitationally bound systems will or will not participate in the presumed expansion of the universe. (They will not, provided that $G*$ behaves so as to satisfy the SEP.)

The Principle of Equivalence fails in the relativistic regime

Einstein's *Gedankenexperiment*

We first consider Einstein's famous thought experiment in which an observer rides a vehicle (rocket ship or elevator) undergoing a constant acceleration, g, in the y-direction in a field-free region of space. We imagine that the vehicle's velocity is zero (at $y = 0$) with respect to an outside observer at time zero. At that instant, a particle moving freely in a straight line (with respect to the outside observer) at a right angle to the line of motion of the accelerating vehicle passes the midline($x = 0$) of that vehicle. Since the vehicle is at rest at $t = 0$, the coordinates are related by the Galilean transformation at that time:

$$x' = x \qquad y' = y - gt^2/2 \qquad t' = t$$

For the particle, $\quad x = vt \quad \Rightarrow \quad x' = vt' \quad$ and $\quad y = 0 \quad \Rightarrow \quad y' = -gt'^2/2$

The path of the particle as seen in the vehicle is simply $y' = -g(x'/v)^2/2$, a parabola. The curvature of the path at $x' = 0$ is $d^2y'/dx'^2 = -g/v^2$, and the magnitude of the radius of curvature is

$$R = v^2/g$$

This expression is supposed to apply to photons as well as massive particles.

Exact solution: particles with rest mass

Now we consider the motion of particles in the Schwarzschild field. The energy E and angular momentum L are conserved quantities. (As usual $f_r = \sqrt{1 - r_S/r}$ where $r_S = 2GM/c^2$)

$$E = \frac{m * c^2}{\sqrt{1-\beta^2}} = \frac{mc^2}{\sqrt{1-\beta^2}} f_r \; ; \; L = \frac{m * v}{\sqrt{1-\beta^2}} r* = \frac{mv}{\sqrt{1-\beta^2}} f_r\left(\frac{r}{f_r}\right) = \frac{mc\beta}{\sqrt{1-\beta^2}} r$$

Since these are constants of the motion, we may evaluate them at the closest point of approach where $r = a$. Also henceforth, $\beta = \beta_{r=a}$

$$E = \frac{mc^2}{\sqrt{1-\beta^2}} f_a \quad \text{and} \quad cL = mc^2 \frac{\beta}{\sqrt{1-\beta^2}} a$$

The analysis is based upon the treatment in Landau & Lifshitz's *Classical Theory of Fields*, eqn. (98.5). (L&L's M is our L.)

$$dr/d\phi = r^2 \sqrt{(E/cL)^2 - [(mc^2/cL)^2 + 1/r^2](1 - r_S/r)}$$

$$\frac{E}{cL} = \frac{f_a}{a\beta} \quad \text{and} \quad \frac{mc^2}{cL} = \frac{\sqrt{1-\beta^2}}{a\beta}$$

$$u^2 = \left(\frac{E}{cL}\right)^2 - \left(\frac{mc^2}{cL}\right)^2 f_r^2 - r^{-2} f_r^2 = f_a^2 (a\beta)^{-2} - (1-\beta^2) f_r^2 (a\beta)^{-2} - r^{-2} f_r^2$$

To determine the curvature we need $\dfrac{d^2 r}{d\phi^2}$

$$\frac{d^2 r}{d\phi^2} = \frac{dr}{d\phi}\frac{d}{dr}\frac{dr}{d\phi} = \frac{dr}{d\phi}\frac{d}{dr}(r^2 u) = \frac{dr}{d\phi}\left[2ru + \frac{r^2}{2u}\frac{d}{dr}(u^2)\right]$$

where we have used the previously defined $dr/d\phi = r^2 u$, and will again below

$$\frac{d^2 r}{d\phi^2} = \frac{dr}{d\phi}\left[\frac{2}{r}\frac{dr}{d\phi} + \frac{r^4}{2}\left(\frac{dr}{d\phi}\right)^{-1}\frac{d}{dr}(u^2)\right]$$

$$\frac{d}{dr}(u^2) = -\frac{1-\beta^2}{\beta^2}a^{-2}(r_S r^{-2}) - (-2r^{-3} - 3r_S r^{-4}) = 2r^{-3} - 3r_S r^{-4} - \frac{1-\beta^2}{\beta^2}r_S a^{-2} r^{-2}$$

$$\frac{d^2 r}{d\phi^2} = \frac{2}{r}\left(\frac{dr}{d\phi}\right)^2 + \frac{r^4}{2}(2r^{-3} - 3r_S r^{-4} - \frac{1-\beta^2}{\beta^2}r_S a^{-2} r^{-2})$$

$$= \frac{2}{r}\left(\frac{dr}{d\phi}\right)^2 + r - \frac{3}{2}r_S - \frac{1-\beta^2}{2\beta^2}r_S a^{-2} r^2$$

The radius of curvature is given by

$$R = \frac{[r^2 + (dr/d\phi)^2]^{3/2}}{r^2 + 2(dr/d\phi)^2 - r(d^2r/d\phi^2)}$$

Inserting the above expression for $\dfrac{d^2r}{d\phi^2}$, the denominator becomes

$$Denom = r^2 - r\,[r - \frac{3}{2}r_s - \frac{1-\beta^2}{2\beta^2}r_s a^{-2} r^2] = \frac{r_s r}{2}\left(3 + \frac{1-\beta^2}{\beta^2}\left(\frac{r}{a}\right)^2\right)$$

Thus

$$R = \frac{[r^2 + (dr/d\phi)^2]^{3/2}}{\dfrac{r_s r}{2}\left(3 + \dfrac{1-\beta^2}{\beta^2}\dfrac{r^2}{a^2}\right)} \qquad \text{Then, since } \left.\frac{dr}{d\phi}\right|_{r=a} = 0,$$

$$R\Big|_{r=a} = \frac{2a^2}{r_s}\frac{\beta^2}{1+2\beta^2} = \frac{2a^2\beta^2}{(2GM/c^2)}\frac{1}{1+2\beta^2} = \frac{v^2}{GM/a^2}\frac{1}{1+2\beta^2} = \frac{v^2}{g}\left(\frac{1}{1+2\beta^2}\right)$$

Clearly, for significantly relativistic motion, this expression differs markedly from that predicted by the principle of equivalence.

Exact solution: photons

For photons, $E = h\nu$ and $L = (h\nu/c)a^* = (h\nu/c)a f_a^{-1}$

Then $\dfrac{E}{cL} = a^{-1}f_a$, and eqn. (98.5), for $m = 0$, becomes

$$\frac{dr}{d\phi} = r^2\sqrt{a^{-2}f_a - r^{-2}f_r^2} = r^2\sqrt{a^{-2}f_a - r^{-2} + r_s r^{-3}} \equiv r^2 u$$

$$\frac{d^2r}{d\phi^2} = \frac{dr}{d\phi}\frac{d}{dr}\frac{dr}{d\phi} = \frac{dr}{d\phi}\frac{d}{dr}[r^2 u] = \frac{dr}{d\phi}\left(2ru + r^2\frac{du}{dr}\right) = \frac{dr}{d\phi}\left(2ru + \frac{r^2}{2u}\frac{d}{dr}(u^2)\right)$$

$$= \frac{dr}{d\phi}\left[\frac{2}{r}\frac{dr}{d\phi} + \frac{r^4}{2}\left(\frac{dr}{d\phi}\right)^{-1}\frac{d}{dr}(u^2)\right] = \frac{2}{r}\left(\frac{dr}{d\phi}\right)^2 + \frac{r^4}{2}\frac{d}{dr}(-r^{-2} + r_s r^{-3})$$

$$= \frac{2}{r}\left(\frac{dr}{d\phi}\right)^2 + r - \frac{3}{2}r_s$$

The denominator in the expression for R is

$$r^2 + 2\left(\frac{dr}{d\phi}\right)^2 - r\frac{d^2r}{d\phi^2} = \frac{3}{2}r_s r$$

Finally, then, the radius of curvature for the photon at $r = a$ is just

$$R = \frac{2a^2}{3r_s} = \frac{1}{3}\left(\frac{a^2}{2GM/c^2}\right) = \frac{1}{3}\frac{c^2}{g}$$

This conforms to the result for massive particles for $v \rightarrow c$.

Thus, whereas Einstein's equivalence principle predicts the radius of curvature to be

$R = v^2/g$ for photons, ($v \rightarrow c$) as well as for particles of any velocity, $v = \beta c$.

In contrast, $R = \dfrac{v^2}{g}\left(\dfrac{1}{1+2\beta^2}\right)$ for particles, while $R = \dfrac{1}{3}\dfrac{c^2}{g}$ for photons.

The reason for the discrepancy lies in the fact that a real gravitational field induces spatial curvature, a phenomenon that Einstein did not anticipate when in 1907 he formulated the principle of equivalence. See Appendix VII, part 3.

Other calculations for phenomena in the Schwarzschild field are presented in appendices. The Einstein effect (the 'bending of starlight') and the Shapiro effect (the delay of electromagnetic signals passing close to a massive body) are treated in Appendix VI. A careful treatment of 'Einstein's elevator' is presented in Appendix VIII.

End of Digressions

Part II. (c) Generalizing the variable rest mass interpretation to stationary spacetimes that are asymptotically static at infinity, with the Kerr field as an example

Introduction

To this point we have considered only the Schwarzschild field, which is not just stationary, but static. The VRM interpretation and the Telemetric system can be extended to fields that are stationary provided that they are asymptotically static at infinity. In non-static stationary fields, the metric tensor contains mixed space-time components which cannot be eliminated except at the cost of introducing explicit time dependence.

The approach taken is to separate dynamical (frame-dragging) effects from those that may be thought of as purely gravitational effects. For each point in the field, a velocity is calculated which maximizes the rate of a clock moving with that velocity. The residual slowing of such clocks is assumed to be purely gravitational and, in accord with the Variable Rest Mass (VRM) Interpretation, the concomitant length dilation effect fixes the relation between the usual geometry defined by proper measurements on the one hand, and the telemetric geometry, on the other.

In order to test the theory, the method is applied to the Kerr solution, and the results are used to calculate the location of the famous unstable photon orbits. The orbits are calculated by assuming that the they will lie on circles whose (telemetric) curvature matches that induced in the ray paths by the shear in the frame-dragging velocity. The results are exactly those calculated directly by the usual methods.

In dealing with non-static but stationary fields, one must disentangle geometric effects from the dynamical effects associated with the so-called 'frame-dragging' phenomenon. The Kerr solution provides an excellent example of this. Here the medium of light transmission seems to be moving in circles in a vortex-like manner as evidenced by the fact that light rays reflected by a series of mirrors arranged on a circle centered on the axis of rotation will require less time when directed with the rotation as compared with rays directed oppositely. Furthermore, in the limit of infinitely many mirrors, i.e., for a circular path, the results are compatible with an interpretation, which assigns an azimuthal velocity, v, to that medium, such that the circuit times are inversely proportional to $c + v$ and $c - v$ respectively. Furthermore, and this is a crucial point, the rate of a clock is maximized when it 'goes with the flow,' i.e., when it moves with the same azimuthal velocity, v. Finally, proper clocks that are fixed in position are slowed by the factor $\sqrt{1 - v^2 / c^2}$ in relation to the local maximum rate. This is understood in terms of the special relativity effect, since these clocks are in relative motion with respect to the flowing aether.

Even the locally fastest, 'go-with-the-flow' clocks run slowly as compared with rate of 'clocks at infinity.' According to the telemetric interpretation, this residual slowing results from the rest-mass reduction effect, and the concomitant length dilation effect fixes the relationship between telemetric lengths and proper lengths in the v-moving frame. These considerations allow one to disentangle the dynamical and geometric effects.

The general stationary field

For the general case, one must determine the velocity, v, at each point in space-time such that the rate of a clock moving with that velocity is a maximum. The metric under consideration is quite general except for the fact that none of the elements of the metric tensor are time dependent. Fields with such a metric is said to be stationary. Explicitly, the metric is $ds^2 = g_{ik}dx^i dx^k$, where i and k range over all time and space coordinates (0, 1, 2, 3). It is useful at times to separate space and time coordinates. In such cases the purely spatial coordinates will use Greek letters for space-indices (1, 2, 3). In this scheme the metric tensor is written as

$$ds^2 = g_{00}(dx^0)^2 + 2g_{0\alpha}dx^0 dx^\alpha + g_{\alpha\beta}dx^\alpha dx^\beta$$

The fastest clocks

The rate of a clock relative to world time x^0, is ds/dx^0 where

$$(ds/dx^0)^2 = g_{00} + 2g_{0\alpha}(dx^\alpha/dx^0) + g_{\alpha\beta}(dx^\alpha/dx^0)(dx^\beta/dx^0)$$

Setting the derivative of $(ds/dx^0)^2$ with respect to (dx^α/dx^0) equal to zero yields the system of equations for the motion (dx^α/dx^0) which will maximize the rate of a clock. The system of equations which results simplifies to

$$g_{\alpha\beta}\, dx^\beta/dx^0 = -g_{0\alpha}$$

The inverse of the pure space submatrix $g_{\alpha\beta}$ is [40]

$$g^{\alpha\beta} - g^{0\alpha}g^{0\beta}/g^{00}$$

The solution to the system is thus

$$dx^\alpha/dx^0 = -(g^{\alpha\beta} - g^{0\alpha}g^{0\beta}/g^{00})g_{0\beta} = g^{0\alpha}/g^{00}$$

Here we have used the trick of expanding summations to all indices and using the identity

$$g^{ik}g_{jk} = \delta^i_j$$

Thus for instance,

$$g^{\alpha\beta}g_{0\beta} = g^{\alpha k}g_{0k} - g^{\alpha 0}g_{00} = \delta^\alpha_0 - g^{\alpha 0}g_{00} = -g^{\alpha 0}g_{00} \quad \text{since } \delta^\alpha_0 = 0.$$

This trick is used repeatedly below.

Inserting this result into the expression for $(ds/dx^0)^2$ one finds

$$(ds/dx^0)^2 = 1/g^{00}$$

---

[40] L.D. Landau and E.M. Lifshitz, *The Classical Theory of Fields*. (Addison-Wesley, Reading, MA, 1962), rev. 2nd ed. This is the contravariant counterpart of eqn. (84.7), p. 273. See also paragraph 2 on page 274.

For the Schwarzschild metric, $1/g^{00} = g_{00}$. Our result thus corresponds to the well-known result for the slowing of clocks, $ds/dx^0 = \sqrt{g_{00}}$, in that field.

The aether flow velocity
The contravariant components of the flow velocity are defined by [41]
$$\beta^\alpha = v^\alpha/c = (dx^\alpha/dx^0)/\{\sqrt{g_{00}}[1+(g_{0\alpha}/g_{00})(dx^\alpha/dx^0)]\}$$, which upon inserting
$$dx^\alpha/dx^0 = g^{0\alpha}/g^{00}$$ the above reduces to
$$\beta^\alpha = \sqrt{g_{00}}\, g^{0\alpha}$$
The covariant components are obtained using the pure space metric tensor [42]
$$\lambda_{\mu\nu} = -g_{\mu\nu} + g_{0\mu}g_{0\nu}/g_{00}$$
Thus
$$\beta_\alpha = \lambda_{\alpha\mu}\beta^\mu = g_{0\alpha}/\sqrt{g_{00}}$$
The physical velocity has magnitude β where
$$\beta^2 = \beta_\alpha\beta^\alpha = g_{0\alpha}g^{0\alpha} = 1 - g_{00}g^{00}$$
and finally, the gamma factor for the flow is
$$\gamma = 1/\sqrt{1-\beta^2} = 1/\sqrt{g_{00}g^{00}}$$

The above derivation of the expression for the flow velocity is perhaps too abstract: consequently, a more intuitive derivation is presented in Appendix IX.

A coordinate system that goes with the flow
Next we shall introduce what may be called 'go-with-the-flow' coordinates $\underline{x}^\alpha$ satisfying
$$d\underline{x}^\alpha = dx^\alpha - (g^{0\alpha}/g^{00})dx^0$$
Inserting these reduces the invariant line element to

$$ds^2 = (1/g^{00})(dx^0)^2 + g_{\alpha\beta}d\underline{x}^\alpha d\underline{x}^\beta$$

Clearly, the rest mass reduction factor (the gravitational potential) is $1/\sqrt{g^{00}}$.

The metric according to telemetric system of measurement
From this we can deduce the telemetric line element, which uses world time, and bases its distance measurements on echo ranging using world time. Thus we set $g*_{00} = 1$. Then

---

[41] Ibid. p. 293.
[42] Ibid. p. 273.

$$\lambda *_{\alpha\beta} = -g^{00}g_{\alpha\beta} \quad \text{and} \quad ds *^2 = (dx^0)^2 + g^{00}g_{\alpha\beta}d\underline{x}^\alpha d\underline{x}^\beta = g^{00}ds^2.$$

Thus the conformal relationship between telemetric and proper metrics holds in the more general case, provided that one properly takes account of the dynamical effects of 'frame dragging' (aether flow).

An apt example: The Kerr field
The Kerr Metric in the usual coordinates can be written[43]

$$ds^2 = \rho^{-2}(\rho^2 - r_S r)(dx^0)^2 + 2\rho^{-2}r_S ra\sin^2\theta\, d\phi\, dx^0$$
$$- \rho^{-2}u^4\sin^2\theta\, d\phi^2 - \Delta^{-1}\rho^2\, dr^2 - \rho^2\, d\theta^2$$

Here we have introduced the expressions
$r_S = 2GM/c^2$ and $a = L/Mc$, where L and M are the angular momentum
and mass of the rotating object, respectively, as well as the usual expressions
$\rho^2 = r^2 + a^2\cos^2\theta$, and $\Delta = r^2 + a^2 - r_S r$
along with a new definition $u^4 = \rho^2(r^2 + a^2) + r_S r\, a^2\sin^2\theta$.

The application of the general results above to the Kerr solution yields, after extensive algebraic manipulation, the following results for the 'flow' velocity and related parameters.

$$d\phi/dx^0 = r_S ra/u^4; \quad \beta = r_S ra\sin\theta/(\rho^2\sqrt{\Delta}); \quad \gamma = (\rho^2\sqrt{\Delta})/[\sqrt{(\rho^2 - r_S r)}\, u^2]$$
$$(d\tau/d\tau*)_{max} = 1/\sqrt{g^{00}} = (\rho\sqrt{\Delta})/u^2$$

This last form is equal to the mass-reduction factor, i.e., for any proper mass, m,

$$m*/m = 1/\sqrt{g^{00}} = (\rho\sqrt{\Delta})/u^2$$

and thus the event horizon occurs where $\Delta = 0$. Thus there are two such surfaces, the 'spheres' (r is only a parameter) defined by

$$r_h = (r_S/2)[1 \pm \sqrt{1 - (2a/r_S)^2}]$$

The region between these horizons seems unphysical, since in that region, the coordinate r becomes timelike while t becomes spacelike. The region for which r falls in the range between zero and the smaller horizon value may represent another universe (Shangri-La?) which seems not to have been investigated. For values of the

---

[43] J M Bardeen "Timelike and Null Geodesics in the Kerr Metric" in *Black Holes,* Les Houches 1972, Gordon and Breach, 1973, p. 219

parameter, a, greater than $r_S/2$, both horizons vanish, and this exotic universe merges with the innerspace already connected to our universe.

Three metrics for the Kerr field
The metric forms of interest are
stationary
$$ds^2 = \rho^{-2}(\rho^2 - r_S r)(dx^0)^2 + 2\rho^{-2} r_S r a \sin^2\theta \, dx^0 d\phi$$
$$- \rho^{-2} u^4 \sin^2\theta \, d\phi^2 - \rho^2 / \Delta \, dr^2 - \rho^2 \, d\theta^2$$

go-with-the-flow
$$ds^2 = u^{-4}\rho^2 \Delta (dx^0)^2 - \rho^{-2} u^4 \sin^2\theta \, d\underline{\phi}^2 - \rho^2 / \Delta \, dr^2 - \rho^2 d\theta^2$$

telemetric
$$ds*^2 = (dx^0)^2 - u^8/(\rho^4\Delta)\sin^2\theta \, d\underline{\phi}^2 - u^4/\Delta^2 \, dr^2 - u^4/\Delta \, d\theta^2$$

The wave nature of matter inferred from the Kerr field
Imagine a test mass dropped from rest at great distance – from infinity, in effect – in the central plane of the Kerr field. The particle will have zero angular momentum, so one might naively imagine that it would fall straight in with a constant azimuthal angle, ϕ.

Alternatively, one might guess that the aether flow would exert a drag on the object, so that its path would be a spiral following, but lagging, the local aether flow. The surprising truth is that the 'particle' exhibits no inertia with respect to the local aether flow, but is quietly swept up by the medium, always matching the local velocity of the aether in the ϕ direction. Thus in the 'go-with-the-flow' system of reference, the object appears to be falling straight in, with no velocity component in the ϕ direction.[44] This behavior is exactly what one would expect of a purely wave-like disturbance, having no being independent of the medium in which it 'waves.'

Proof of the pudding: the photon orbits in the Kerr field
In order to demonstrate that the the VRM interpretation is correct, we shall investigate the two unstable photon orbits which occur in the Kerr solution. The orbits will be calculated by assuming that the orbits will lie on circles whose telemetric curvature matches the curvature induced in the ray paths by the shear in the 'frame dragging' velocity. The results conform exactly those calculated directly by the usual methods. We need first to define the functions $r*$ and $R*$. Let

$$r* = u^4/(\rho^2\sqrt{\Delta}) \quad \text{and} \quad dR*/dr = u^2/\Delta,$$

---

[44] Ibid., p.221.

so that the element of telemetric length in the ϕ and r directions are $r * \sin\theta\, d\phi$, and $dR*$, respectively. Having defined $r*$, we digress momentarily to consider the stenosurface(s), surface of minimum (or maximum) circumference surrounding the rotating object (star). The stenosurface is determined by the equation $d/dr(r*)=0$, which leads to

$$2r^7 - 3r_S r^6 + 2(1+2C^2)a^2 r^5 - (3+2C^2)r_S a^2 r^4$$
$$+[(4C^2+2C^4)a^2 + 3S^2 r_S^2]a^2 r^3 - (C^4+2S^4)r_S a^4 r^2$$
$$+(2C^2 a^2 - S^2 r_S^2)C^2 a^4 r + (1+S^2)C^2 r_S a^6 = 0$$

where $C = \cos\theta$ and $S = \sin\theta$. In the section, $\theta = \pi/2$, the above reduces to a polynomial form that is only fifth degree in r and just quadratic in a^2, which may then be solved for a^2. The result is, setting $x = r/r_S$ and $A = (2a/r_S)$

$$A^2 = x[2x^2 - 3x + 3 \pm \sqrt{4x^4 + 4x^3 - 3x^2 - 18x + 9}]$$

This expression gives all of the critical surfaces: for the outermost of these, which corresponds to the stenosurface, use x in the range (1.5 , 1.2737) with the negative sign, for A in the range (0 , 1.757). For values of $A > 1.757$, use the positive sign and $x > 1.2737$. Appendix X presents an indirect procedure for finding a section of the stenosurface of the Kerr solution for any polar angle.

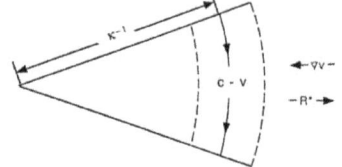

figure 1. Curvature of rays induced by shear in the aether flow velocity

Returning to the photon orbits, one would expect that in the outer region, beyond the stenosurface, wavefronts propagating in the counter-flow direction would be turned inward because of the shear in the flow. The curvature, κ_r, of the ray is easily calculated. From figure 1, one has two equal expressions for the angle

$$[c - v(R*-\Delta R*)]\Delta t/(\kappa_r^{-1} - \Delta R*) = [c - v(R*+\Delta R*)]\Delta t/(\kappa_r^{-1} + \Delta R*)$$

Expanding v in a Taylor series, one has for $\Delta R* \to 0$,

$$\kappa_r = -(dv/dR*)/(c-v) = d/dR * [\ln(c-v)]$$

On the other hand, referring to figure 2, one has

$$\rho*/r* = dR*/dr*$$

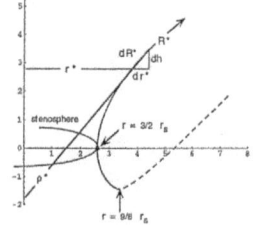

figure 2. Embedding diagram showing the sensible radius $\rho*$ (scale in units of r_S)

whence the telemetric curvature of a circle in the equatorial section is

$$\kappa_o = 1/\rho^* = (1/r^*)(dr^*/dR^*) = d/dR^*[\ln(r^*)]$$

If the telemetric picture correctly deals with dynamics and geometry, then the photon orbit should occur when these curvatures match, implying

$$d/dR^*[r^*/(c-v)] = 0$$

Now $2\pi r^*/(c-v)$ is just the telemetric transit time for a light signal traversing the ring of mirrors, and one would expect that this circle would correspond to an orbit when the transit time is a minimum. But expectations aside, it turns out that the above equation is in fact satisfied by the actual photon orbit as determined in the usual way, as will now be shown.

For the region inside the stenosurface, a wavefront propagating in the same direction as the rotation will be turned so as to circle that surface. In this case the curvatures are

$$\kappa_r = d/dR^*[\ln(c+v)] \quad \text{and} \quad \kappa_o = d/dR^*[\ln(r^*)]$$

so that the orbit should satisfy $d/dR^*[r^*/(c+v)] = 0$

The two cases will be treated simultaneously. With the benefit of hindsight, one may simplify the work by writing

$$r^*/(1 \mp \beta) = r^*(1 \pm \beta)/(1 - \beta^2) = (\rho^2\sqrt{\Delta} \pm r_S r a \sin\theta)/(\rho^2 - r_S r)$$

The orbits lie in the equatorial plane, so that $\rho = r$. Also, R^* is a monotone function of r. We may thus write the equations as

$$d/dr[(r\sqrt{\Delta} \pm r_S a)/(r - r_S)] = 0$$

This results in the following equation, valid for both orbits $(\pm)$

$$4r^3 - 12r_S r^2 + 9r_S^2 r - 8r_S a^2 = 0$$

Setting $A = 2a/r_S$, the solutions may be written, first for the outer, retrograde orbit,

$$r_{ph-}/r_S = 1 + \cos(2/3 \cos^{-1}(A)) \quad \text{for } A < 1$$
$$= 1 + \cosh(2/3 \cosh^{-1}(A)) \quad \text{for } A > 1$$

68

and for the "inner" (smaller values of r), prograde orbit,

$$r_{ph+} / r_S = 1 + \cos(2/3 \cos^{-1}(-A))$$

These are the well-known (unstable) photon orbits.[45] The third root,

$$r_{ph+??} / r_S = 1 - \cos(1/3 \cos^{-1}(1 - 2A^2))$$

is a prograde orbit in the 'Shangri-La' universe mentioned earlier.

Figure 3 shows the embedding diagram for the equatorial section of the Kerr geometry for three values of the rotation parameter, $A = 2a/r_S$. The dots indicate the location of the photon orbits.

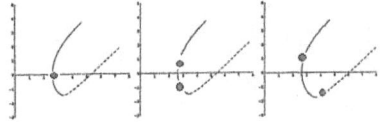

It is to be hoped that this analysis will serve to establish the fact that the aether is a substance apart from the geometry, and that it behaves rather like an incompressible liquid capable of flowing.

fig. 3. Telemetric imbedding diagrams for the equatorial section of the Kerr field for three values of the rotation parameter ($A \approx 0.0, 0.2, 0.4$), showing the location of the unstable photon orbits. Scale in units of r_g.

The massless Kerr solution

There is an interesting sidelight to the Kerr solution, namely the case in which the mass parameter M is taken to be zero while, nevertheless, what is thought of as the rotation parameter, a, remains finite. This is treated in Appendix XI.

---

[45] Ibid., p.224.

Part III. Cosmological Implications of Variable Rest Masses

Introduction
In Cosmology, the VRM interpretation implies an unexpected and even startling re-interpretation of the cosmological red shift, identifying the function $A(t)$ as the 'mass evolution function,' instead of the function describing the expanding scale of the universe. The interpretation does not involve the rather mysterious concept of the expansion of space itself. Perhaps the best features are those that are not present, namely, the infinite densities and singularity of the big bang, and the faster-than-light expansion of space itself, which, under the standard interpretation, is an accepted feature of inflationary theories.

Momentum conservation implies variable rest masses
For the solutions of Einstein's equations thought to be descriptive of the expanding universe, typified by the Robertson Walker cosmology, spatial homogeneity is assumed. Noether's theorem [46] in this case requires that momentum be conserved. Since the momentum of a photon is simply $h\nu/c$, this requirement (assuming h and c to be constants) would, in particular, mean that the photons of the cosmic microwave background (CMB) field have maintained their original frequency unchanged in all the billions of years they have been in flight since their birth at recombination time. But that seems incredible on the face of it! Those photons were originally emitted in the infrared range (peak wavelength ~ 10 microns) by a 3000°K hydrogen-helium plasma: how can they have maintained their frequency if they now appear as black body radiation with a peak wavelength 1000 times greater, near 10 millimeters? The only possible answer is that in the intervening time the characteristic frequencies of all processes, including our frequency 'standards,' have increased by a factor of one thousand!

One way that this might have happened is if the rest mass of all material bodies were to have increased by a factor of one thousand since the recombination era. For if this were the case, the characteristic frequencies of our measuring devices would have increased one thousand fold, explaining the apparent red shift. This remarkable possibility is not unreasonable: in fact, it is demanded by the same theorem of Noether. In an addendum below I derive an integral of motion for a test mass moving through the galaxies of a Robertson Walker universe, namely,

$$A(t)\,\beta\,/\,\sqrt{1-\beta^2} = \text{constant.}$$

---

[46] C Lanczos, *The Variational Principles of Mechanics*, 4th ed., Dover (1986), pp.401-05. See also the excellent article on Noether in *Wikipedia*.

Here $\beta = v/c$, is the relativity parameter, and $A(t)$ is the scale function describing the manner in which, according to the conventional interpretation, the universe expands with the time. Clearly, since $A(t)$ increases with time, the relativity parameter, β, must decrease. However, since the momentum is given by $p = mc\beta / \sqrt{1 - \beta^2}$, this would seem to be inconsistent with conservation of momentum – unless, that is, the rest mass of the particle were to increase in proportion to the function $A(t)$. To meet the demand of Noether's theorem, we define the 'true' rest mass of the test particle whose proper rest mass is m, as

$$m^* = m A(t) / A_{pe},$$

where A_{pe} is the value of A at some fixed reference time, say, that of the present epoch. Multiplying the integral of motion by the constant mc / A_{pe}, we have

$$m^* c \beta / \sqrt{1 - \beta^2} = p^* = \text{constant}.$$

This satisfies the demand that momentum be conserved. Not only that, but the increase in rest masses also explains the cosmological red shift: as rest masses increase, the characteristic frequency of all systems – including those used to measure the frequency of the CMB – increase in proportion.

The evolution of the universe: The equations for $A(t)$

Here we investigate the cosmological implications of the VRM interpretation by developing equations for the function $A(t)$, here interpreted as describing the manner in which rest mass evolves, increasing with the passage of time. The analysis is based upon the conservation of momentum, which is implied by the assumption of the large scale homogeneity of the universe.

Preliminaries

Let $a = A(t)/\tilde{A}$ in which ~ *denotes* the present epoch, and let $u = \beta/\sqrt{1-\beta^2}$.

From the preceding section,

$$p* = m*c\beta/\sqrt{1-\beta^2} = mc\,(A/\tilde{A})\,\beta/\sqrt{1-\beta^2} = mc\,au = mc\tilde{u} = const.$$

Then $E* = m*c^2/\sqrt{1-\beta^2} = mc^2\,a\sqrt{1+u^2} = mc^2\sqrt{a^2+(au)^2} = mc^2\sqrt{a^2+\tilde{u}^2}$

Some kinetic theory

For the moment we suppress the asterisks. Consider a cubical box, L on a side. When a particle bounces off the side normal to the direction x, the momentum imparted is equal to twice the x-component of the particles momentum:

$\Delta p = 2mc\beta_x/\sqrt{1-\beta^2}$ the time between hits on this wall is

$$\Delta t = 2L/c\beta_x \,, \text{ so the force imparted is } f_x = \Delta p/\Delta t = mc^2\beta_x^2/L\sqrt{1-\beta^2}$$

Now $\beta^2 = \beta_x^2 + \beta_y^2 + \beta_z^2$ and for the isotropic case assumed, $\beta_x^2 = \frac{1}{3}\beta^2$. Thus

$$f_x = \frac{mc^2\beta^2}{3L\sqrt{1-\beta^2}} \quad \text{The pressure is then} \quad \wp = \frac{f_x}{L^2} = \frac{mc^2\beta^2}{3L^3\sqrt{1-\beta^2}} = \frac{mc^2\beta^2}{3V\sqrt{1-\beta^2}}$$

The pressure due to N particles is N times this, and the number density is $n = N/V$

Thus finally we have $\wp = \dfrac{nmc^2\beta^2}{3\sqrt{1-\beta^2}}$.   Now $\dfrac{\beta^2}{\sqrt{1-\beta^2}} = \dfrac{u^2}{1+u^2}\sqrt{1+u^2} = \dfrac{u^2}{\sqrt{1+u^2}}$

Restoring the asterisks and tildes, $u = \tilde{u}/a$ and $\dfrac{\beta^2}{\sqrt{1-\beta^2}} = \dfrac{\tilde{u}^2}{a\sqrt{a^2+\tilde{u}^2}}$ and the

pressure is

$$\wp* = \frac{nm*c^2\beta^2}{3\sqrt{1-\beta^2}} = \frac{a\,nmc^2}{3}\frac{\tilde{u}^2}{a\sqrt{a^2+\tilde{u}^2}} = \frac{nmc^2}{3}\frac{\tilde{u}^2}{\sqrt{a^2+\tilde{u}^2}}$$

And the Energy density is

$$\varepsilon^* = \frac{nm^*c^2}{\sqrt{1-\beta^2}} = anmc^2\sqrt{1+u^2} = nmc^2\sqrt{a^2+\tilde{u}^2}$$

The energy-momentum tensor T_{ik} is diagonal with

$$T_{00} = \varepsilon^*, \quad T_{kk} = -\wp^* \quad for \ k = 1,2,3$$

We shall limit our consideration to the case of a single field with rest mass. Also, for the sake of simplicity, and since space appears to be Euclidean, the metric is assumed to be

$$ds^2 = a^2(x^0)\ [(dx^0)^2 - (dx^1)^2 - (dx^2)^2 - (dx^3)^2]$$

Calculating the Einstein tensor
The metric tensor is diagonal with

$$g_{00} = a^2, \quad g_{kk} = -a^2; \quad g^{00} = a^{-2}, \quad g^{kk} = -a^{-2}$$

The only non-zero Christoffel symbols are

$$\Gamma^0_{00} = \tfrac{1}{2}g^{00}\frac{\partial g_{00}}{\partial x^0} = \tfrac{1}{2}a^{-2}(a^2\dot{}) = \frac{\dot{a}}{a}$$

$$\Gamma^k_{k0} = \tfrac{1}{2}g^{kk}\frac{\partial g_{kk}}{\partial x^0} = \tfrac{1}{2}(-a^{-2})(-a^2\dot{}) = \frac{\dot{a}}{a}$$

$$\Gamma^0_{kk} = \tfrac{1}{2}g^{00}(-\frac{\partial g_{kk}}{\partial x^0}) = \tfrac{1}{2}(a^{-2})(a^2\dot{}) = \frac{\dot{a}}{a}$$

$$g = \det(g_{**}) = -a^8, \qquad \sqrt{-g} = a^4$$

The Ricci tensor is may be expressed as

$$R_{ik} = 1/\sqrt{-g}\,\frac{\partial}{\partial x^l}(\Gamma^l_{ik}\sqrt{-g}) - \frac{\partial^2}{\partial x^i \partial x^k}[\ln(\sqrt{-g})] - \Gamma^m_{il}\Gamma^l_{km} \qquad \text{For the case at hand,}$$

$$R_{ik} = a^{-4}\frac{\partial}{\partial x^l}(\Gamma^l_{ik}a^4) - \frac{\partial^2}{\partial x^i \partial x^k}[\ln a^4] - \Gamma^m_{il}\Gamma^l_{km} = 4(\dot{a}/a)\Gamma^0_{ik} + \frac{\partial}{\partial x^0}\Gamma^0_{ik} - 4\delta_{0i}\delta_{0k}\frac{\partial}{\partial x^0}(\dot{a}/a) - R^{(3}_{ik}$$

Where the third terms, $R^{(3)}_{ik} = \Gamma^m_{il}\Gamma^l_{km}$, will be evaluated below.

For $i \neq k$, $\Gamma^0_{ik} = 0$, and

$$\boxed{R_{ik} = -R^{(3)}_{ik}}$$

For $i = k = 0$

$$R_{00} = 4(\dot{a}/a)\Gamma_{00}^0 + \frac{\partial}{\partial x^0}\Gamma_{00}^0 - 4\frac{\partial}{\partial x^0}(\dot{a}/a) - R_{00}^{(3)} = 4(\dot{a}/a)^2 - 3[\ddot{a}/a - (\dot{a}/a)^2] - R_{00}^{(3)}$$

$$\boxed{R_{00} = 7(\dot{a}/a)^2 - 3\ddot{a}/a - R_{00}^{(3)}}$$

For $i = k \neq 0$

$$R_{kk} = 4(\dot{a}/a)\Gamma_{kk}^0 + \frac{\partial}{\partial x^0}\Gamma_{kk}^0 - R_{kk}^{(3)} = 4(\dot{a}/a)^2 + \frac{\partial}{\partial x^0}(\dot{a}/a) - R_{kk}^{(3)} = 3(\dot{a}/a)^2 + \ddot{a}/a - R_{kk}^{(3)}$$

$$\boxed{R_{kk} = 3(\dot{a}/a)^2 + \ddot{a}/a - R_{kk}^{(3)}}$$

Now regarding the products of Christoffel symbols, $R_{ik}^{(3)} = \Gamma_{il}^m \Gamma_{km}^l$

$$R_{0k}^{(3)} = \Gamma_{0l}^m \Gamma_{km}^l = \Gamma_{00}^0 \Gamma_{k0}^0 + \Gamma_{0k}^0 \Gamma_{k0}^k + \Gamma_{00}^k \Gamma_{kk}^0 + \Gamma_{0k}^k \Gamma_{kk}^k$$
$$(l,m) \qquad 0,0 \qquad k,0 \qquad 0,k \qquad k,k$$

$$\boxed{R_{0k}^{(3)} = 0} \quad \text{since} \quad \Gamma_{k0}^0 = \Gamma_{0k}^0 = \Gamma_{00}^k = \Gamma_{kk}^k = 0$$

$$R_{00}^{(3)} = \Gamma_{0l}^m \Gamma_{0m}^l = \Gamma_{00}^0 \Gamma_{00}^0 + \Gamma_{0j}^0 \Gamma_{00}^j + \Gamma_{00}^j \Gamma_{0j}^0 + \Gamma_{0j}^j \Gamma_{0j}^j$$
$$(l,m) \qquad 0,0 \qquad j,0 \qquad 0,j \qquad j,j$$

Now $\Gamma_{0j}^0 = \Gamma_{00}^j = 0$, so

$$R_{00}^{(3)} = \Gamma_{00}^0 \Gamma_{00}^0 + \sum_{j=1,2,3}\Gamma_{0j}^j \Gamma_{0j}^j = 4(\dot{a}/a)^2$$

$$\boxed{R_{00}^{(3)} = 4(\dot{a}/a)^2}$$

$$R_{kk}^{(3)} = \Gamma_{kl}^m \Gamma_{km}^l = \Gamma_{k0}^0 \Gamma_{k0}^0 + \Gamma_{kk}^0 \Gamma_{k0}^k + \Gamma_{k0}^k \Gamma_{kk}^0 + \Gamma_{kk}^k \Gamma_{kk}^k$$
$$(l,m) \qquad 0,0 \qquad k,0 \qquad 0,k \qquad k,k$$

Now $\Gamma_{k0}^0 = \Gamma_{kk}^k = 0$, so

$$R_{kk}^{(3)} = 2\Gamma_{kk}^0 \Gamma_{k0}^k = 2(\dot{a}/a)^2$$

$$\boxed{R_{kk}^{(3)} = 2(\dot{a}/a)^2}$$

Collecting the above results, we have $R_{0k} = 0$ and for the diagonal components,

$$R_{00} = -3\frac{\ddot{a}}{a} + 3\frac{\dot{a}^2}{a^2}$$ and $$R_{kk} = \frac{\ddot{a}}{a} + \frac{\dot{a}^2}{a^2}$$ for $k = 1, 2, 3$

The scalar 'curvature', R, is given by

$$R = g^{ik}R_{ik} = g^{ii}R_{ii} = g^{00}R_{00} + 3g^{kk}R_{kk} = a^{-2}\left[-3\frac{\ddot{a}}{a} + 3\frac{\dot{a}^2}{a^2}\right] - 3a^{-2}\left[\frac{\ddot{a}}{a} + \frac{\dot{a}^2}{a^2}\right] = -6a^{-2}\frac{\ddot{a}}{a}$$

The Einstein tensor $G_{ik} = R_{ik} - \frac{1}{2}g_{ik}R$ is diagonal with components

$$G_{00} = 3\frac{\dot{a}^2}{a^2}$$ and $$G_{kk} = -2\frac{\ddot{a}}{a} + \frac{\dot{a}^2}{a^2}$$ for $k = 1,2,3$

The gravitational field equations $G_{ik} = \frac{8\pi G}{c^4}T_{ik}$, are then

$$3\frac{\dot{a}^2}{a^2} = \frac{8\pi G}{c^4}nmc^2\sqrt{a^2 + \tilde{u}^2}$$ and $$-2\frac{\ddot{a}}{a} + \frac{\dot{a}^2}{a^2} = -\frac{8\pi G}{3c^4}nmc^2\frac{\tilde{u}^2}{\sqrt{a^2 + \tilde{u}^2}}$$

Introduce a new time variable $\tau = \Omega\,(x^0/c)$, where $\Omega = \sqrt{\frac{8}{3}\pi G nm}$. The equations may be written as (reusing the dot)

$$\frac{\dot{a}^2}{a^2} = \sqrt{a^2 + \tilde{u}^2}$$ and $$2\frac{\ddot{a}}{a} - \frac{\dot{a}^2}{a^2} = \frac{\tilde{u}^2}{\sqrt{a^2 + \tilde{u}^2}}$$

The product of the right hand members of the two equations is a constant, $\tilde{u}^2$.

Thus the product of the members of the left hand side, $\frac{\dot{a}^2}{a^2}\left[2\frac{\ddot{a}}{a} - \frac{\dot{a}^2}{a^2}\right]$, must also be a constant. The form of this expression suggests an exponential function, easily guessed to be $\exp\left(\sqrt{\tilde{u}}\,\tau\right)$. Indeed, for

$$a(\tau) = \exp\left(\sqrt{\tilde{u}}\,\tau\right), \quad \frac{\dot{a}^2}{a^2}\left[2\frac{\ddot{a}}{a} - \frac{\dot{a}^2}{a^2}\right] = \tilde{u}^2$$

Since this holds for all times we shall be satisfied that the rest mass evolution in the universe is well described by

$$a(\tau) \approx \exp\left(\sqrt{\tilde{u}}\,\tau\right)$$

Note that $a \to 0$ *only for* $\tau \to -\infty$. Many theologians will not be pleased with this result.

To estimate the frequency $\Omega = \sqrt{\tfrac{4}{3}\pi G n m}$, we have, for an assumed mass density of $3x10^{-27} kg/m^3$, $\Omega \approx 1.295 x10^{-18} \sec^{-1} \approx 4.087 x10^{-11} year^{-1}, \Omega^{-1} \approx 2.447 x10^{10} years$. Let's introduce T as the time as kept by a physical clock, i.e., T corresponds to 'proper' time, running forward from the 'big bang.' From the assumed metric,

$$dT = a(t)\,dt = \exp\!\left(\sqrt{\tilde{u}}\ \Omega\, t\right) dt$$

$$T = \int_{-\infty}^{t} \exp(\sqrt{\tilde{u}}\ \Omega t')\,dt' = \left(\sqrt{\tilde{u}}\ \Omega\right)^{-1} \exp(\sqrt{\tilde{u}}\ \Omega\, t)$$

The proper time $\tilde{T}$ from the 'big bang' at $T = 0$ to the present epoch at $t = 0$ is just

$$\boxed{\tilde{T} = \frac{1}{\sqrt{\tilde{u}}}\,\Omega^{-1}}$$

Regarding $\tilde{u} = \tilde{\beta}/\sqrt{1-\tilde{\beta}^2}$, there is great uncertainty, so we consider a range of values

$$\sqrt{\tilde{u}} = \tfrac{1}{2},\ 1,\ 2 \quad \Rightarrow \quad \tilde{\beta} = 1/\sqrt{1+\sqrt{\tilde{u}}^{\,-4}} = 0.2425,\ 0.7071,\ 0.9701$$

For the 'present age of the universe', $\tilde{T}$, we have, for $\sqrt{\tilde{u}} = \tfrac{1}{2},\ 1,\ 2$,

$$\tilde{T} = 4.895 x10^{10}\ years,\ 2.447 x10^{10}\ years,\ 1.224 x10^{10}\ years$$

The highly relativistic estimate for $\sqrt{\tilde{u}}$ fits well with the conventional figure of $1.375 x10^{10}\ years$, which would require $\sqrt{\tilde{u}} = 1.780 \quad \Rightarrow \quad \tilde{\beta} = 0.732$.

Note that $t = \dfrac{1}{\sqrt{\tilde{u}}}\Omega^{-1}\ln(\sqrt{\tilde{u}}\ \Omega\, T)$ so

$$a = \exp\left(\sqrt{\tilde{u}}\ \Omega\, t\right) = \exp\left[\sqrt{\tilde{u}}\ \Omega \left(\frac{1}{\sqrt{\tilde{u}}}\Omega^{-1}\left(\ln \sqrt{\tilde{u}}\ \Omega\, T\right)\right)\right] = \sqrt{\tilde{u}}\ \Omega\, T$$

Thus in terms of the proper time, T, rest masses simply increase linearly with time:

$$\boxed{a(T) = \sqrt{\tilde{u}}\ \Omega\, T}.$$

Momentum conservation of itself implies that A(t) is an exponential function of world time

Upon careful reflection, it can be shown that the fact that the evolution of rest mass is an exponential function of world time follows as a direct consequence of momentum conservation. Consider a source and two observers lying on a straight line at distances d_1 and $d_2 > d_1$ from the source. For brevity, let $\omega = \sqrt{\tilde{u}}\ \Omega$. Generally, $\lambda_{obs} = \lambda_{std}\ a\,(\omega t_{obs})/a\,(\omega t_{emit})$. Let λ_1 & λ_2 be the wavelengths observed at d_1 & d_2

$$\lambda_2\,a\,(\omega t_{emit}) = \lambda_{std}\,a\,(\omega t_2\,) = \lambda_{std}a[\omega(t_1) + \omega(d_2 - d_1)/c]$$

Momentum conservation implies that wavelengths of light remain constant in traveling between any two points. If that is so, the spectra measured at d_2 will differ from that observed at d_1 only as a result of the changes in the instruments at d_2 caused by the increase in rest masses during the time interval $(d_2 - d_1)/c$. And of course, the same is true for the time interval $t_1 = d_1/c$. Thus the function on the right hand side above must be representable as the product of some function of d_1/c and the same function of $(d_2 - d_1)/c$. Thus

$$a\,[d_1/c + (d_2 - d_1)/c] = f(d_1/c)\cdot f\,[(d_2 - d_1)/c]$$

But when $d_2 = d_1$, we must have $f = a$, and thus the function a satisfies the functional equation

$$a(x+y) = a(x)\cdot a(y),$$

Only the exponential function satisfies this equation. Thus we conclude that

$$a(t) \approx \exp[\omega(t - \tilde{t}\,)],$$

Photons are not stretched as space itself expands: They were born red shifted

According to the VRM interpretation, the red shift has nothing to do with the recession of distant galaxies: it is simply a consequence of the fact that the frequency of any given atomic transition was, in past eras, reduced as compared with the present values for the same transition.

The behavior of the effective temperature of the particles whose momentum is conserved is treated below. The early universe was somewhat hotter than it is presently, but the temperature at the instant of the 'big bang' was still finite.

To put it mildly, the cosmology being proposed will be considered to be highly speculative. But it's based firmly upon Noether's theorem, which may not be dismissed lightly.

According to the VRM interpretation, the function, $A(t)$, should be called the 'mass evolution function,' rather than the 'distance scale function.' Whatever it's called, it is not understood. No one really knows why the universe should expand – or, more properly, why space itself should expand – nor am I able to propose any explanation of why rest masses should evolve, increasing with time.

Addendum: An integral of motion in a Robertson Walker universe

The Robertson Walker metric may be written [cf. Landau and Lifshitz, *The Classical Theory of Fields*, rev. 2nd ed., (Pergamon Press, 1962, p. 380)

$$ds^2 = c^2 dt^2 - A^2(t) \, [d\chi^2 + S^2(\chi) \, (d\theta^2 + \sin^2 \theta \, d\phi^2)]$$

where the function $S(\chi)$ is $\sin \chi$, χ, or $\sinh \chi$ depending on whether the curvature of the universe is positive, zero, or negative; corresponding, respectively, to a closed and finite universe, a Euclidean and infinite universe, or open and infinite universe. Introducing a new time variable, η, defined by $cdt = Ad\eta$, the metric becomes

$$ds^2 = A^2(\eta)[d\eta^2 - d\chi^2 - S^2(\chi)(d\theta^2 + \sin^2 \theta \, d\phi^2)]$$

Without loss of generality, one may consider motion in the χ direction, for which the geodesic equation is

$$d^2\chi/ds^2 + \Gamma^\chi_{\eta\eta}(d\eta/ds)^2 + 2\Gamma^\chi_{\eta\chi}(d\eta/ds)(d\chi/ds) + \Gamma^\chi_{\chi\chi}(d\chi/ds)^2 = 0$$

But $\Gamma^\chi_{\eta\eta} = 0$, $\Gamma^\chi_{\eta\chi} = A^{-1}(dA/d\eta)$, and $\Gamma^\chi_{\chi\chi} = 0$. Thus the geodesic equation reduces to

$$d^2\chi/ds^2 + 2\,A^{-1}(dA/d\eta)(d\eta/ds)(d\chi/ds) = 0$$

Inserting $(dA/d\eta)(d\eta/ds) = dA/ds$ and integrating we have $A^2(d\chi/ds) =$ constant.

The proper velocity, v, is defined by $v = dl/dt$, where $dl = Ad\chi$ for the motion considered, and $dt = c^{-1}Ad\eta$. Thus, $v/c = d\chi/d\eta$, and the integral may be written

$$A^2(d\chi/ds) = A^2(d\chi/d\eta)/(ds/d\eta) = A^2(v/c)/(ds/d\eta) = \text{constant}$$

But directly from the line element, for the motion considered, one has

$$ds/d\eta = A\sqrt{1 - (d\chi/d\eta)^2} = A\sqrt{1 - (v/c)^2}$$

Inserting this, and introducing $\beta = v/c$, the integral of motion may be written

$$A(\eta) \, \beta/\sqrt{1 - \beta^2} = \text{constant}.$$

How temperature varies in the universe according to the variable rest mass hypothesis.

Assume that $T \approx \dfrac{2}{3}\dfrac{kinetic\ energy}{k_B} = \dfrac{2}{3}\dfrac{mc^2}{k_B}\left[\dfrac{1}{\sqrt{1-\beta^2}}-1\right].$

Define $\theta = \dfrac{3}{2}\dfrac{k_B T}{mc^2}$ i.e., $\theta = \left[\dfrac{1}{\sqrt{1-\beta^2}}-1\right]$

We have seen that momentum conservation in a homogeneous Robertson Walker universe requires

$$\frac{m^* c\beta}{\sqrt{1-\beta^2}} = \frac{\alpha\, mc\,\beta}{\sqrt{1-\beta^2}} = constant = \frac{mc\beta_{pe}}{\sqrt{1-\beta_{pe}^2}} \quad where \quad \alpha = \frac{A(t)}{A_{pe}}$$

So we have $\qquad \alpha\dfrac{\beta}{\sqrt{1-\beta^2}} = \dfrac{\beta_{pe}}{\sqrt{1-\beta_{pe}^2}}$

Now

$$\frac{\beta}{\sqrt{1-\beta^2}} = \sqrt{\frac{1}{1-\beta^2}-1} = \sqrt{\left[\frac{1}{\sqrt{1-\beta^2}}-1\right]\left[\frac{1}{\sqrt{1-\beta^2}}-1+2\right]} = \sqrt{\theta(\theta+2)}$$

Thus the integral of motion can be written as

$$\alpha\sqrt{\theta(\theta+2)} = \sqrt{\theta_{pe}(\theta_{pe}+2)} \quad or \quad \theta^2 + 2\theta - \alpha^{-2}\theta_{pe}(\theta_{pe}+2) = 0$$

$$\theta = \sqrt{1+\alpha^{-2}\theta_{pe}(\theta_{pe}+2)}-1$$

What we want is

$$T^* = \frac{2}{3}\frac{m^* c^2}{k_B}\left[\frac{1}{\sqrt{1-\beta^2}}-1\right] = \frac{2}{3}\frac{mc^2}{k_B}\alpha\theta$$

$\theta*$ vs α for θ_{pe} = 1.0, 0.8, 0.6, 0.4, 0.2, & 0.05

$$\theta^* = \alpha\theta = \sqrt{\theta_{pe}(\theta_{pe}+2)+\alpha^2}-\alpha \quad and \quad \theta^*\big|_{\alpha=0} = \sqrt{\theta_{pe}(\theta_{pe}+2)}$$

The temperature of the universe was somewhat higher at earlier epochs, but remains finite even at the instant of the 'big bang.' In fact, from the previous section

$$\tilde{\beta} = \beta_{pe} = 0.734 \;\Rightarrow\; \theta_{pe} = 0.46777 \;\Rightarrow\; \theta^*\big|_{\alpha=0} = 1.0744 = 2.297\,\theta_{pe}.$$

Gravitationally bound bodies in a Robertson Walker universe

Often people wonder whether the expansion of space implies that the distance between the Sun and the planets would increase as $A(t)$ increases.

For simplicity we consider the case of two bodies in circular orbit. The governing (Newtonian) equations

$$m_1 r_1 \omega^2 = G m_1 m_2 / (r_1 + r_2)^2 \quad \text{and} \quad m_2 r_2 \omega^2 = G m_1 m_2 / (r_1 + r_2)^2$$

may be combined to yield

$$G = (r_1 + r_2)^3 \omega^2 / (m_1 + m_2)$$

Meanwhile, the total angular momentum, J, can be expressed as follows

$$J = m_1 r_1^2 \omega + m_2 r_2^2 \omega = m_1 r_1 \omega^2 (r_1 / \omega) + m_2 r_2 \omega^2 (r_2 / \omega)$$

$$= G m_1 m_2 / [(r_1 + r_2) \omega]$$

$$= [(r_1 + r_2)^3 \omega^2 / (m_1 + m_2)] \, [m_1 m_2 / [(r_1 + r_2) \omega]]$$

$$= [m_1 m_2 / (m_1 + m_2)] (r_1 + r_2)^2 \omega$$

Note that this expression does not involve G directly. Thus the dependence of the angular momentum upon $A(t)$, the 'scale factor' (or mass evolution function), of the universe is simply determined by the dimensionality of angular momentum in terms of mass, length and time.

Now in an isotropic universe, angular momentum is conserved: This means that J does **not** depend upon $A(t)$: that is,

$$J * / J \sim A^0$$

The equation above then requires

$$(m) + 2(l) - (t) = 0$$

But according to the analysis that led to the variable rest mass interpretation

$$(l) = -(1/3)[2 + (m)] \quad \text{and} \quad (t) = -1.$$

Thus the invariance of angular momentum requires

$$(m) - 2(1/3)[2 + (m)] - (-1) = 0 \quad \Rightarrow \quad (m) = 1$$

Inserting this result into the expression $G = (r_1 + r_2)^3 \omega^2 /(m_1 + m_2)$, we see that,

writing $\quad G*/G \sim A^{(G)}$

$$(G) = 3(l) - 2(t) - (m) = -[2 + (m)] - 2[-1] - (m) = -2$$

The original equations may be cast in the form

$$r_1 + r_2 = [(m_1 + m_2)/(m_1 m_2)^2][J^2/G]$$

$$\omega = [(m_1 m_2)^3 /(m_1 + m_2)][G^2 / J^3]$$

In obvious notation one finds that

$$(r) = -3(m) - (G) + 2(J) = -1$$

$$(\omega) = 5(m) + 2(G) - 3(J) = 1$$

Thus the orbital parameters behave exactly as do measuring rods and real clocks, so that the **measured** orbital radius and orbital period will not be affected by the increase in $A(t)$, regardless of whether it expresses the supposed expansion of the universe or the mass evolution function.

On the other hand, if one were to imagine that $G*$ was in fact an actual constant, one would find that $(r) = -3$ and $(\omega) = 5$. Remembering that $(l) = -1$ and $(t) = -1$, this would lead to **observable** secular changes in orbital radius and frequency expressed by $(r) = -2$ and $(\omega) = 4$. (Notice that the orbital radius would appear to **decrease** rather than increase, contrary to what one would be expected if space were expand everywhere.)

Taking the Hubble constant to be

$$H = \frac{\dot{A}}{A} \cong 71.9 \,(km/s)/Mp\sec = 73.5 \, x \, 10^{-12}/\,year,$$

one would expect the changes in measured orbital distance, r, and period, P, to be

$$\frac{\dot{r}}{r} = -2\frac{\dot{A}}{A} = -2H \approx -1.47 \, x \, 10^{-10}/\,year \quad \& \quad \frac{\dot{P}}{P} = -4H \approx -2.94 \, x \, 10^{-10}/\,year \cdot$$

Such secular changes would probably be undetectable amongst the many other perturbations.

Part IV. Other variable rest mass theories

Dicke

Robert Dicke formulated the Strong Equivalence Principle (SEP) in the form used in this book. He realized that time and length intervals must vary together, inversely as rest mass varies, if the SEP is to hold. He investigated a situation in which the mass scaling factor was a definite function of a postulated scalar field, developing a hybrid theory called the Brans-Dicke Scalar-Tensor theory,[47] by including scalar field terms in the Lagrangian of general relativity. Though he did not acknowledge it immediately, he also realized that the gravitational 'constant', G, must vary oppositely as does mass and by twice the factor. He compared a selected version of his theory to the usual solar system tests of GR (excepting the Shapiro delay of signals passing close to planets or the sun). Ironically, he missed the opportunity to understand the whole thing immediately because of his misunderstanding of the gravitational red shift. Dicke[48] asserted that

> "[The gravitational red shift] is not a test of anything but energy conservation and mass-energy equivalence."

This same misunderstanding blinded him to the possibility of breaking what appears to be a global conformal symmetry, according to which none of the changes allowable under the SEP could be remotely detected. Quoting Dicke:[49]

> "Imagine, if you will, that you are told by a space traveller that a hydrogen atom on Sirius has the same diameter as one on Earth. A few moments thought will convince you that the statement is either a definition or else meaningless."

Einstein's wonderfully simple 1911 argument proving that in a stationary field the red shift must be understood as a reduction of spectral lines at the source also establishes the fact that freely moving light rays travel with constant frequency. This, in turn, permits the comparison of the spectral frequencies of remote sources to the corresponding spectra of terrestrial sources. Clearly, this is equivalent to measuring the value of the Rydberg period, $h^3/(m_e e^4)$, at the remote source, which, after a bit of thinking, tells us the value of the electron rest mass at the remote location. Finally

---

[47] Brans, C. H.; Dicke, R. H. (1961). "Mach's Principle and a Relativistic Theory of Gravitation". *Physical Review* **124** (3): 925–935.

[48] Dicke R H (1965) *The Theoretical Significance of Experimental Relativity*, Gordon and Breach, New York, p.25

[49] Ibid, p.35

then, applying the SEP tells us all we need to know regarding the diameter of the hydrogen atom.

Bekenstein

Jacob Bekenstein [50] investigates the possibility that the rest masses of particles are determined by a universal field, which he calls the rest-mass field. He even assumes that the field is represented by a tensor, so as to allow for the possibility of non-isotropic inertial mass. Following Dicke, he develops the theory by incorporating rest-mass field terms in the Lagrangian of General Relativity. Finally, he applies his theory to the several solar system tests of the theory in order to answer his basic question "Do rest masses vary or not?"

But this whole endeavor is really unnecessary, since the proof of variable rest masses follows from the simplest examination of the gravitational red shift, which must be understood as a reduction of spectral lines at the source. Again, the argument may be bolstered by a quotation from Einstein himself: [51]

> "...we can regard an atom which is emitting spectral lines as a clock, so that the following statement will hold: ***An atom absorbs or emits light of a frequency which is dependent on the potential of the gravitational field in which it is situated.***" [Emphasis Einstein's]

And, as I have shown, this frequency dependence is a direct consequence of the dependence of rest mass on the gravitational potential. So Beckenstein's monumental effort seems not to be relevant to the present study.

Froedge

Froedge [52] assumes that "the total energy of a particle is localized in the volume of the particle, and that there is no energy in the fields related to the particle." This assumption, Froedge believes, is not compatible with 'GR' (Einstein's theory of gravity.) Thus he believes it necessary to develop an alternative theory, taken, it would seem for simplicity sake, to be a scalar theory. He correctly recognizes that the gravitational reduction of rest masses causes a reduction of emission frequencies, thus explaining the red shift, and he adopts a constant time scale, according to which the frequency of a photon rising in a gravitational field is constant.

But Froedge is not aware of the fact that variable rest mass causes not only a reduction of emission frequencies, but also a dilation of every object with rest mass, including measuring rods. It follows that he must accept Einstein's assumption that

---

[50] Bekenstein J D (1977) *Phys Rev D* vol.15, No. 6, pp 1458-1468

[51] A Einstein, *Relativity*, (Crown Publishers, New York, 1952), 15th ed., Appendix III(c), pp.130-131 in hardcover, p. 149 in paper.

[52] Froedge D T (2007) 'Scalar Gravitational Theory with Variable Rest Mass, APS meeting, April 14-17, 2007

the speed of light is reduced in a gravitational field. In fact, he cites the Shapiro effect in support of this assumption.

It is remarkable that Froedge did not make a systematic investigation of the implications of rest mass reduction, through which he would have discovered gravitational length dilation. Equally remarkable is his assumption that the localization of potential energy is incompatible with Einstein's theory of gravity. It is clear that the weight of authority and consensus in physics powerfully inhibits any criticism of the accepted canon, even as regards alternative interpretations of existing mathematical theories, in this case Einstein's theory of gravity.

Hoyle and Narlikar

The theory considered here evolved out of the Steady State Cosmology, which postulated a continuous creation of matter. After several modifications, Hoyle and Narlikar developed a theory in which the particle number is constant, but in which particle masses may vary with position in spacetime. It is a conformally invariant theory of gravity featuring variable particle rest masses.[53] The theory is based upon an action at a distance interaction between  masses, and is Machian in that all particles serve as sources for a scalar 'mass function,' the local value of which, in turn, determines a particle's mass. The Hoyle-Narlikar cosmologies bear an eerie similarity to the VRM cosmology presented in this book – the cosmic red shift is attributed to the epochal increase of the mass function, rather than to an expansion of space.

Narlikar and Halton Arp[54] elaborated on the theory, proposing that the mass function may exhibit departures from a smooth decrease with time and distance, exhibiting regions in which rest masses are very much smaller than the values typical for sources at same distance. Matter in such regions would exhibit anomalously high red shifts. Arp had compiled a catalog of peculiar galaxies, and later found that many quasars, point-like sources exhibiting very large red shifts, were very often associated with these peculiar galaxies, which themselves did  not possess large red shifts. The red shifts of these quasars have been interpreted by mainstream astronomers as indicating great remoteness, and consequently the objects

---

[53] Hoyle & Narlikar, *Proc. R. Soc.* **A282**, 191 (1964), Hoyle & Narlikar, *Proc. R. Soc.* **A294**, 138 (1966) Narlikar J V, *Ann. Physics*, **107**, 325-336, (1977); Narlikar, J V, *Introduction to Cosmology*, 2nd ed. Cambridge University Press 1993

[54] Narlikar and Arp, *ApJ*, **405**, 51-56, (1993)

84

are believed to radiate at unaccountable power levels. Arp disputes the idea that red shifts are always correlated with distance, proposing that newly created matter appears initially with very low rest mass, and consequently has an anomalously high red shift. Narlikar and Arp consider that surfaces may exist on which the mass function takes on a zero value. Near such a region, they argue, quantum fluctuation will be sufficient to cause the production of new particles, resulting in a sort of mini-big bang, as it were.

Sonego

Sebastiano Sonego [55] develops a modified Lorentz-Dirac equation (for the motion of a charged particle under the influence of the electromagnetic self interaction force, F_a, as well as an external electromagnetic field, F_{ab}) which is covariant under conformal transformation $\tilde{g}_{ab} \to e^{-2\Phi} g_{ab}$. He first proves that the electromagnetic self force F_a transforms with weight –1, i.e., $\tilde{F}_a = e^{\Phi} F_a$, as is required for conformal covariance. This alone is not sufficient, however, if one assumes that rest mass is strictly constant. Sonego observes that because of the equivalence of mass and energy, rest mass must transform in the same manner as electromagnetic energy, which has conformal weight -1.[56] Thus $\tilde{m} = e^{\Phi} m$. With this result in hand, Sonego develops a modified Lorentz-Dirac equation that is indeed conformally covariant. The modified equation is

$$v^b \nabla_b p_a = -\nabla_a m + e F_{ab} v^b + F_a$$

The important connection with the theory presented in this book is the identification of the term $-\nabla_a m$ with Newtonian gravitational

force. Sonego suggests that one might well dispense with the notion of gravity, and think only in terms of a mass field. This is quite like the notion put forward in this book that the action of gravity is to transform rest mass energy into kinetic energy and vice versa. The mass field simply replaces, or is identified with, the gravitational field.

Sonego notes that for a neutral particle the equation above, written in terms of $\tilde{g}_{ab}$, reduces to

---

[55] Sebastiano Sonego (1999) *J. Math. Phys.*, vol.40. No.8, 3918-3924

[56] The energy of the electric field due to a charge, e, in the region outside a sphere of radius, r, is $u = e^2/(2r)$, which, since $\tilde{e} = e$, will transform inversely as does length. Thus $\tilde{u} = e^{\Phi} u$.

$$\tilde{v}^b\tilde{\nabla}_b\tilde{p}_a = -\tilde{\nabla}_a\tilde{m}$$

multiplying each member by $\tilde{m}$ one has

$$\tilde{p}^b\tilde{\nabla}_b\tilde{p}_a = -\tilde{\nabla}_a(\tilde{m}^2/2)$$

Thus it is clear that massive particles are affected by the force $-\tilde{\nabla}_a\tilde{m}$, while massless ones simply move on null geodesics. He concludes that gravity appears as just one manifestation of mass, and therefore, massless particles are not affected by it.

Finally, Sonego speculates that there are essentially two contributions to the mass $\tilde{m}$ of a particle; one, small by comparison, due to local matter, that is position dependent and thus gives rise to gravitational forces; and, in contrast, a much larger contribution due to large scale distribution of matter contributing the constant part of the particle mass which is identified with the inertia of the particle. Allow me to say that it is gratifying to note that Sonego's analysis indicates that massless particles are not affected by gravitation forces.

Comment on Machian 'Mass Fields'

The Machian schemes proposed by Hoyle and Narlikar, and by Sonego, in which distant masses create a 'mass-field' that determines the larger, constant part of a particle's rest mass, is seemingly undermined by the fact that the action of nearby masses is to *reduce* the rest mass of the particle.

A rather fanciful alternative is suggested by the standing wave representation of matter. This would require the acceptance of the possibility that space may have a character that is altered by the presence of matter. For instance, in the modeling of 'point' particles, a standing wave in a rolled-up extra dimension, w, was posited. If one imagines that gravitating bodies cause a local increase in the circumference of the w space, and if the standing wave is imagined to expand so as to preserve the number of nodes, then the particle, converting wave energy to kinetic energy, would accelerate in the direction of increasing circumference. In the process the frequency, and hence the rest mass, would be reduced. If W is the circumference of the extra dimension, and if the nodal number is conserved, we can write

$$k_w W = nodal\ number = k_{w\infty}W_\infty, \text{ whence } \omega = ck_w = ck_{w\infty}\,W_\infty/W = \omega_\infty\,W_\infty/W$$

and finally, $m = h\omega/c^2 = m_\infty\,W_\infty/W$

Additional remarks regarding conformal invariance

A conformal transformation, unlike a mere coordinate transformation, actually alters the character of the space (or space-time) under consideration. In ordinary geometry, such transformations may dilate, twist and turn, but angles are preserved unchanged. In relativity, the ratios of length and time intervals are kept invariant, and thus a conformal transformation will leave the speed of light unchanged and keep the light cone invariant.

By conventional definition, a conformal transformation involves a twice differentiable and non-vanishing function, Ω, of the coordinates producing a new metric, $ds*^2 = \Omega^2 ds^2$. Einstein's field equations are not generally conformally invariant, since under conformal transformation the curvature, R, changes to $R* = \Omega^{-2}[R + 6\Delta_4\Omega/\Omega]$, where Δ_4 is the d'Alembertian evaluated with respect to the original metric.[57]

Invariance in physics also requires that the relevant equations remain invariant. There is no problem with Maxwell's equations, but a superficial examination of the Lorentz force equation seems to demand that rest masses must transform inversely as do time and space intervals, as is easily shown. (In fact, if one introduces the Bohr radius and the Rydberg period as paradigms for length and time intervals, as we have done, the invariance of the Lorentz equation is automatic, even without insisting on the invariance of the fundamental electric charge.)

Several studies have suggested that Einstein's Theory of gravity should be general-ized so as to be invariant under arbitrary conformal transformation.[58]
These studies all recognize the additional requirement that rest masses must vary inversely as length and time intervals in order to insure that the equation of physics remain invariant. But the VRM interpretation shows that the relevant equations are already invariant in Einstein's theory of gravity, provided only that the variability of rest masses is accepted. This brings into question the desirability of altering the theory to make it conformally invariant. To do so would in effect declare that all conformally equivalent metrics actually describe the same spacetime, whereas this is not the case, since, as has been noted, the curvature tensor is not invariant under conformal transformation. In fact, the ability to carry out - and *understand* - remote observations of spectra breaks the conformal symmetry that seems to obtain when only local measurements are taken into account. Thus Dicke was not correct in his statement regarding the impossibility of comparing the diameter of a hydrogen atom on Sirius with one on Earth.

---

[57] Narlikar, J V, *Introduction to Cosmology*, 2nd ed., Cambridge University Press (1993), p.266.

[58] R L Ingraham (1952) "Conformal Relativity" *Proc Natl Acad Sci* v38 (10); C Brans, R H Dicke (1961) "Mach's Principle and a Relativistic Theory Of Gravitation". *Physical Review* **124**: 925; R L Dicke (1962) "Mach's Principle And Invariance Under Transformation Of Units, "*Phys. Rev.* **125**, 2163; T Fulton, F Rohrlich and L Witten, "Conformal Invariance in Physics," *Rev. Mod. Phys.*, **34**, (3), 442-457, (1962a); F Hoyle & J V Narlikar, *Proc. R. Soc.* A, **294**, 138-148, (1966).

It is the fact that electromagnetic signals move with constant frequency in stationary fields that enables the remote intercomparison of the energy levels of atoms of the same type. This, along with the Dicke's Strong Equivalence Principle, permits the remote inter-comparison of masses as well as the diameter of atoms.

As previously noted, neither Newton's nor Einstein's theories are conformally invariant so long as the gravitational constant, G, is taken to be an actual constant. In the Schwarzschild field conformality requires that $G^* = f^{-2}G$. The same is true of Einstein's cosmical constant, Λ : conformality requires $\Lambda^* = f^{-2}\Lambda$.

Appendices

Appendix I. A formal derivation of the Lorentz transformation

The keys to the derivation are these:

(1) The transformation T must be symmetric in the sense that $T^{-1}(v) = T(-v)$
(2) The speed of light must be invariant.

We begin by pointing out that the transformation must be linear in the coordinates, since the choice of origin must be arbitrary. We expect that T will have the following form

$$ct' = \gamma_0[ct - u(\beta)x] \quad x' = \gamma_1[x - \beta ct] \quad y' = \gamma_2 y \quad z' = \gamma_2 z$$

First, we have assumed that the primed system, K', moves in the $+x$ direction with speed $v = \beta c$. [Regarding (1) above, note that $T^{-1}(v)$ and $T(-v)$ both describe how K' would observe K moving off in the $-x$ direction.] Second, we have introduced $u(\beta)$ anticipating that a special synchronization must be implemented if the speed of light is to be constant in the K' system. Third, we have assumed that measurements made in the y and z direction will be affected, if at all, in the same way.

It is convenient to introduce a bit of matrix notation. We can write T in matrix form as $X' = TX$ with inverse $X = T^{-1}X'$

$$\begin{bmatrix} ct' \\ x' \\ y' \\ z' \end{bmatrix} = \begin{bmatrix} \gamma_0 & -\gamma_0 u & 0 & 0 \\ -\gamma_1\beta & \gamma_1 & 0 & 0 \\ 0 & 0 & \gamma_2 & 0 \\ 0 & 0 & 0 & \gamma_2 \end{bmatrix} \begin{bmatrix} ct \\ x \\ y \\ z \end{bmatrix} \qquad T^{-1} = \begin{bmatrix} \gamma_1/d & \gamma_0 u/d & 0 & 0 \\ \gamma_1\beta/d & \gamma_0/d & 0 & 0 \\ 0 & 0 & \gamma_2^{-1} & 0 \\ 0 & 0 & 0 & \gamma_2^{-1} \end{bmatrix}$$

Where $d = \gamma_0\gamma_1(1 - \beta u)$

Clearly the value of γ_2 cannot be altered if the direction of motion of K' were to be reversed: hence γ_2 must be an even function of β. This together with the require-ment $T^{-1}(v) = T(-v)$, implies that $\gamma_2 = 1$.

Note that the determinant of the upper 2x2 submatrix of $T^{-1}(v)$ equals d^{-1}, while that of the upper submatrix of $T(-v)$ equals d. Thus $T^{-1}(v) = T(-v)$ demands that $d = 1$. This and another appeal to symmetry requires that

90

$$\begin{bmatrix} \gamma_1 & \gamma_0 u(\beta) \\ \gamma_1 \beta & \gamma_0 \end{bmatrix} = \begin{bmatrix} \gamma_0 & -\gamma_0 u(-\beta) \\ \gamma_1 \beta & \gamma_1 \end{bmatrix}$$

Thus, $\gamma_0 = \gamma_1 = \gamma$, and $u(-\beta) = -u(\beta)$, i.e., u is an odd function of β.

Finally, we demand that the speed of light be constant: $\{x = ct\} \Rightarrow \{x' = ct'\}$

$$x' - ct' = \gamma(x - \beta ct) - \gamma(ct - u(\beta)x) = \gamma[(x - ct) + (u(\beta)x - \beta ct)] = \gamma[(u(\beta) - \beta]x$$

So finally, it is determined that $u(\beta) = \beta$. Going back to the requirement that the determinant of the submatrix must be unity, i.e., that $d = 1$, we find

$$d = \gamma_0 \gamma_1 (1 - \beta u) = \gamma^2 [1 - \beta^2] \quad \Rightarrow \quad \gamma = 1/\sqrt{1 - \beta^2}$$

Then we can write

$$T = \begin{bmatrix} \gamma & -\gamma\beta & 0 & 0 \\ -\gamma\beta & \gamma & 0 & 0 \\ 0 & 0 & 1 & 0 \\ 0 & 0 & 0 & 1 \end{bmatrix}$$

Or in the usual format

$$ct' = \gamma[ct - \beta x] \quad x' = \gamma[x - \beta ct] \quad y' = y \quad z' = z$$

An Historical Note: Voigt's transformation
There is an odd and ironical twist to the history of this transformation. In 1887, the very year of the Michelson Morley experiment, Waldemar Voigt, a Göttingen mathematical physicist (an expert in crystallography) published a paper on the Doppler Effect. In the paper he presented a transformation under which Maxwell's equations were invariant. Voigt's transformation differed from the Lorentz transformation by the conformal factor, γ^{-1}.

We label the transformations 'V' for Voigt, and 'L' for Lorentz

$$
V: \begin{bmatrix} ct' \\ x' \\ y' \\ z' \end{bmatrix} = \begin{bmatrix} 1 & -\beta & 0 & 0 \\ -\beta & 1 & 0 & 0 \\ 0 & 0 & \gamma^{-1} & 0 \\ 0 & 0 & 0 & \gamma^{-1} \end{bmatrix} \begin{bmatrix} ct \\ x \\ y \\ z \end{bmatrix} = \gamma^{-1} \begin{bmatrix} \gamma & -\gamma\beta & 0 & 0 \\ -\gamma\beta & \gamma & 0 & 0 \\ 0 & 0 & 1 & 0 \\ 0 & 0 & 0 & 1 \end{bmatrix} \begin{bmatrix} ct \\ x \\ y \\ z \end{bmatrix} = \gamma^{-1} L
$$

So Voigt's transformation is a simple conformal modification of the Lorentz's. It follows that the inverse is just γ times the inverse of the Lorentz transformation, and hence it is not true that V^{-1} is just V with $-\beta$ replacing β: so it fails the symmetry requirement.

Regarding the rate of a moving clock: For x' fixed,

$$ x = \beta ct \quad \text{and} \quad ct' = ct - \beta x = ct - \beta(\beta ct) = (1 - \beta^2)ct = \gamma^{-2}ct . $$

Thus moving clocks are slowed by *two* factors of γ^{-1}. This makes sense since the transverse length of objects is increased by the factor γ, while the component of velocity of light in the transverse direction is reduced to $\sqrt{c^2 - v^2} = \gamma^{-1}c$.

Evidently, Voigt's motive was merely to find a transformation that would leave invariant the source-free Maxwell equations, and, clearly, changing scale by a constant factor will have no effect. Perhaps the most remarkable aspect of Voigt's paper was the introduction of what came to be called by Lorentz, the 'local time.' The introduction of 'local time' is equivalent to the Einstein scheme for the synchronization of clocks, which, as I have argued, is the 'stealth postulate' that makes possible the universal constancy of the speed of light. Lorentz and Voigt were in communication, and it is possible that Lorentz was inspired to introduce 'local time' by Voigt's example.

Appendix II. Stellar Aberration and the Doppler Effect

Aberration

An observer moving with respect to the aether at speed v encounters an incoming light ray that crosses the observer's line of motion at true angle θ (as determined by an observer at rest in the aether). A co-moving observer, magically not subject to shrinking measuring rods, would see the ray crossing at angle ϕ. The following relation would pertain:

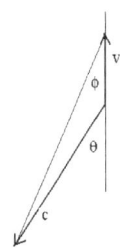

or

$$\frac{v}{\sin(\theta - \phi)} = \frac{c}{\sin \phi} \quad \text{whence} \quad \sin \theta \cos \phi - \cos \theta \sin \phi = \frac{v}{c} \sin \phi$$

$$\tan \phi = \frac{\sin \theta}{\cos \theta + v/c}$$

But the actual observer overestimates distances along the direction of his motion and therefore will underestimate the tangent of an angle by the factor $\sqrt{1 - v^2/c^2}$. His measured angle will thus be

$$\bullet \quad \tan \phi' = \sqrt{1 - v^2/c^2} \left[\frac{\sin \theta}{\cos \theta + v/c} \right] = \frac{\sqrt{1 - \beta^2} \sin \theta}{\cos \theta + \beta}$$

Often one needs the sine and cosine functions:

$$\sin \phi' = \tan \phi' \cos \phi' = \frac{\tan \phi'}{\sqrt{1 + \tan^2 \phi'}} = \frac{\sqrt{1 - \beta^2} \sin \theta}{(\cos \theta + \beta) \sqrt{1 + \frac{(1 - \beta^2) \sin^2 \theta}{(\cos \theta + \beta)^2}}}$$

$$= \frac{\sqrt{1 - \beta^2} \sin \theta}{\sqrt{(\cos \theta + \beta)^2 + (1 - \beta^2) \sin^2 \theta}} = \frac{\sqrt{1 - \beta^2} \sin \theta}{\sqrt{\cos^2 \theta + 2\beta \cos \theta + \beta^2 + \sin^2 \theta - \beta^2 \sin^2 \theta}}$$

$$= \frac{\sqrt{1 - \beta^2} \sin \theta}{\sqrt{1 + 2\beta \cos \theta + \beta^2 - \beta^2 \sin^2 \theta}} = \frac{\sqrt{1 - \beta^2} \sin \theta}{1 + \beta \cos \theta}$$

$$\bullet \quad \sin \phi' = \frac{\sqrt{1 - \beta^2} \sin \theta}{1 + \beta \cos \theta}$$

$$\cos\theta = \sqrt{1 - \sin^2\theta} = \frac{\sqrt{(1 + \beta\cos\theta)^2 - (1 - \beta^2)\sin^2\theta}}{1 + \beta\cos\theta}$$

$$= \frac{\sqrt{1 + 2\beta\cos\theta + \beta^2\cos^2\theta - \sin^2\theta + \beta^2\sin^2\theta}}{1 + \beta\cos\theta}$$

$$= \frac{\sqrt{\cos^2\theta + 2\beta\cos\theta + \beta^2}}{1 + \beta\cos\theta} = \frac{\cos\theta + \beta}{1 + \beta\cos\theta}$$

- $$\cos\phi' = \frac{\cos\theta + \beta}{1 + \beta\cos\theta}$$

Doppler Effect

$$dt = \frac{1}{\sqrt{1 - v^2/c}\ f_0}$$

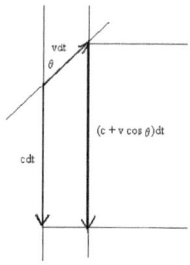

For a distant emitter of base frequency f_0 moving away at speed v at an angle θ to the line of sight, the distance between successive wavecrests will be, accounting for the Larmor reduction of frequency,

$$\lambda = \frac{(c + v\cos\theta)}{\sqrt{1 - v^2/c^2}\ f_0}.$$ And the observed frequency

$$f = c/\lambda = \left[\frac{\sqrt{1 - v^2/c^2}}{1 + (v/c)\cos\theta}\right] f_0$$

Appendix III. The Cosmic Flash *Gedankenexperiment*

Here we imagine a flashbulb emitting a pulse of light that propagates in all directions at the speed of light with respect to the aether frame, K. (It is not assumed that the flashbulb itself is at rest in the aether.) A system of reference, K', moving at velocity, v, observes the flash. One point in K', labeled A, is coincident with the origin of the flash. At some predetermined time, t'_A, after the flash, every observer in system K' who

detects the flash, instantly sends out a secondary signal. These signals are judged in K' to be emitted simultaneously at time t'_A. We seek to describe what observers in the aether system, K, will observe regarding these secondary signals sent out from the system K'. As we shall show, what will be observed is a wave of signals sweeping across a prolate ellipsoid in the same direction as the system K' is moving. The following analysis is referenced to the aether system, K.

What we seek to do discover is how the events which make up a 'simultaneous' observation of the outgoing 'sphere' according to observers in a moving system K' will be recorded by observers in the aether frame.

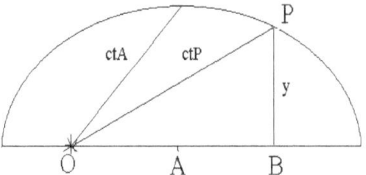

The 'origin' of the flash in the K' system moves along the $+x$ axis from the flash point, O, to point A in time t_A:

$$OA = vt_A = \beta ct_A$$

Event P represents the detection of the outgoing wave by an observer with coordinates (x, y) relative to point A. Point P is directly above point B, i.e., $x = AB$. The events A and P are 'simultaneous:' i.e., the 'Wave of Simultaneity' connects them. We are considering the situation from the standpoint of the rest (aether) system, with respect to which the wave moves to the right at speed, $W_S = c^2/v = c/\beta$. Noting that events P and B are simultaneous in the aether system K, we can write

$$AB = W_S(t_B - t_A) = \beta^{-1}(ct_P - ct_A)$$

Now

$$(ct_P)^2 = (OA + AB)^2 + y^2 = [\beta ct_A + \beta^{-1}(ct_P - ct_A)]^2 + y^2$$

To suppress clutter, let $u = ct_P$ and $\theta = ct_A$

$$u^2 = [\beta\theta + \beta^{-1}(u-\theta)]^2 + y^2 = \beta^2\theta^2 + 2(u-\theta) + \beta^{-2}(u-\theta)^2 + y^2$$

$$\beta^2 u^2 = \beta^4\theta^2 + 2\beta^2(u-\theta)\theta + (u-\theta)^2 + \beta^2 y^2$$

$$(1-\beta^2)u^2 - 2(1-\beta^2)\theta u + (1-2\beta^2 + \beta^4)\theta^2 + \beta^2 y^2 = 0$$

$$(1-\beta^2)(u^2 - 2\theta u + \theta^2) - (1-\beta^2)\theta^2 + (1-\beta^2)\theta^2 + \beta^2 y^2 = 0$$

$$(u-\theta)^2 - \beta^2\theta^2 + [\beta^2/(1-\beta^2)]y^2 = 0$$

$$(u-\theta)^2 + [\beta^2/(1-\beta^2)]y^2 = \beta^2\theta^2$$

$$\left[\frac{u-\theta}{\beta\theta}\right]^2 + \left[\frac{y}{\sqrt{1-\beta^2}\,\theta}\right]^2 = 1$$

Now $\beta^{-1}(u-\theta) = AB = x(P)$ relative to point A at time t_A. Thus we find that the figure swept out by the flash of simultaneity is a simple ellipse with semi-major axis along the direction of motion equal to c times the time elapsed since the flash occurred, and semi-minor axis equal to $\sqrt{1-\beta^2}$ times the semi-major axis. The location of the flash is at the focus of the ellipse, and the eccentricity is equal to β .

$$\left[\frac{x}{ct_A}\right]^2 + \left[\frac{y}{\sqrt{1-\beta^2}\,ct_A}\right]^2 = 1$$

In the figure to the right we've set the scale to ct_A in order to illustrate the proportions of the figure swept out by the flash of 'simultaneity' sent out by observers in K' who are imagined to reside on a sphere (as seen in K') centered on point A.

It is worthwhile to see how this figure is related to the actual location of the 'flashers' (as determined by simultaneous observation in K). All that is required is to 'compress' the figure by accounting for the distance traveled at speed v during the time required for the wave of simultaneity to go from A to the point P in question. Take A's coordinates to be $(0,0)$, and P's to be (x,y). The ordinate, y, is unaffected, of course.

The time required for the wave of simultaneity to reach P is just $\dfrac{x}{W_S} = \beta\dfrac{x}{c}$ and in

that time A will move toward P, at speed $v = \beta c$, the distance

$$v\frac{x}{W_S} = (\beta c)\left(\beta\frac{x}{c}\right) = \beta^2 x \ .$$

The actual position of P relative to A is thus $\left((1-\beta^2)x, y\right)$. The figure observed in K is thus an oblate ellipsoid, in which the diameter in the direction of motion is shortened relative to the diameter in the transverse direction by the FitzGerald contraction factor, $\sqrt{1-\beta^2}$.

Of course, observers in K′ , whose measuring rods contract along the direction of motion, will measure the diameters to be the same and equal to $\sqrt{1-\beta^2}\,ct_A$. The actual diameter of the 'sphere' observed in systems moving through the aether decreases with increasing speed. However, because of clock slowing, $t'_A = \sqrt{1-\beta^2}\,t_A$, and the diameter measured in K′ is just ct'_A, as it should be.

Incidentally, Doppler shifts will be detected in every system except that in which the flashbulb was at rest. Thus such observers will know that the actual source is no longer at the center of the expanding sphere which their measurements detect.

Appendix IV. The Clock on a String *Gedankenexperiment*

This imaginary experiment attempts to measure the velocity of the apparatus with respect to the aether. We imagine a clock being whirled around a central point of the lab system which is moving with speed $v = \beta c$ with respect to the aether.

The basic idea is that the speed of the clock with respect to the aether frame varies between $r\omega - v$ to $r\omega + v$ as the clock circles. The rate of the clock will vary and it is hoped that this variation will be detectable by an observer at the center. Of course, it turns out that no effect can be detected.

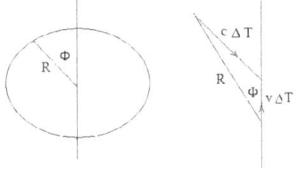

Capitals refer to measurements made in the aether rest system, lowercase to those made in the moving system.

$$R\cos\Phi = \sqrt{1-\beta^2}\,r\cos\omega t \quad \text{and} \quad R\sin\Phi = r\sin\omega t$$

$$R = r\sqrt{1-\beta^2\cos^2\omega t} \qquad \cos\Phi = \frac{\sqrt{1-\beta^2}\,\cos\omega t}{\sqrt{1-\beta^2\cos^2\omega t}} \qquad \sin\Phi = \frac{\sin\omega t}{\sqrt{1-\beta^2\cos^2\omega t}}$$

Let $\Delta S = c\Delta T$ so $v\Delta T = \beta\Delta S$ Then $\Delta S^2 = R^2 + \beta^2\Delta S^2 - 2R\beta\Delta S\cos\Phi$

$$(1-\beta^2)(\Delta S/R)^2 + 2\beta\cos\Phi(\Delta S/R) - 1 = 0$$

$$(1-\beta^2)(\Delta S/R) = \sqrt{\beta^2\cos^2\Phi + (1-\beta^2)} - \beta\cos\Phi = \sqrt{1-\beta^2\sin^2\Phi} - \beta\cos\Phi$$

$$= \frac{(1-\beta^2)}{\sqrt{1-\beta^2\sin^2\Phi} + \beta\cos\Phi}$$

$$\Delta S = \frac{R}{\sqrt{1-\beta^2\sin^2\Phi} + \beta\cos\Phi}$$

Now $1-\beta^2\sin^2\Phi = 1 - \dfrac{\beta^2\sin^2\omega t}{1-\beta^2\cos^2\omega t} = \dfrac{1-\beta^2}{1-\beta^2\cos^2\omega t}$ and

$$\beta\cos\Phi = \frac{\sqrt{1-\beta^2}\,\beta\cos\omega t}{\sqrt{1-\beta^2\cos^2\omega t}}$$

$$\sqrt{1-\beta^2\sin^2\Phi} + \beta\cos\Phi = \frac{\sqrt{1-\beta^2}}{\sqrt{1-\beta^2\cos^2\omega t}}(1+\beta\cos\omega t)$$

$$\Delta S = \frac{r(1-\beta^2\cos^2\omega t)}{\sqrt{1-\beta^2}(1+\beta\cos\omega t)} = \frac{r}{\sqrt{1-\beta^2}}(1-\beta\cos\omega t) = \gamma\, r\,(1-\beta\cos\omega t)$$

Next we calculate the time of reception at the center of a signal from the clock sent at time, t_e. Using the Lorentz transformation for time, we have for the corresponding time in the aether frame

$$T_e = \gamma(t_e + \frac{v}{c^2}r\cos\omega t_e) = \gamma(t_e + (r/c)\beta\cos\omega t_e)$$

The time of flight for the signal is $\Delta S/c = \gamma(r/c)(1-\beta\cos\omega t_e)$ and thus the time of arrival is

$$T_a = T_e + \Delta S/c = \gamma(t_e + (r/c)\beta\cos\omega t_e) + \gamma(r/c)(1-\beta\cos\omega t_e) = \gamma(t_e + r/c)$$

Finally, the time of receipt of the signal by the co-moving observers (with their slow clocks) will be

$$t_a = \gamma^{-1} T_a = t_e + r/c$$

$$t_a - t_e = r/c$$

Thus co-movers will see no modulation of the frequency of signals sent from the clock, and the scheme would fail to detect motion with respect to the aether.

Appendix V. Photon in a superconducting box

In what follows, proper measurements are written plain, while T measurements have an asterisk appended. Recall that the T-length of an object is dilated by the factor $f^{-1} = 1/\sqrt{1 - 2GM/c^2 r}$ in the Schwarzschild field: that is, $L* = f^{-1}L$. Also, the world time, t, is a kosher T-variable, even though we'll not burden it with an asterisk. As a box of proper width, L, is raised, its sides move together with relativity parameter

$$\beta = \frac{1}{c}\frac{d}{dt}\left(\frac{L*}{2}\right) = \frac{1}{2c}\frac{dr}{dt}\frac{d}{dr}(f^{-1}L) = \frac{1}{2}\frac{L}{c}\frac{f'}{f^2}\frac{dr}{dt}$$

The number of bounces suffered by a photon as the box moves upward through distance Δr is

$$N = \frac{\Delta r/(dr/dt)}{(L*/c)} = \frac{c}{L}f\left(\frac{L}{2\beta c}\frac{f'}{f^2}\right)\Delta r = \frac{1}{2\beta}\frac{f'}{f}\Delta r$$

With each bounce the photon energy, $E*$, increases by the factor

$$\left(\sqrt{\frac{1+\beta}{1-\beta}}\right)^2 = \frac{1+\beta}{1-\beta} \quad \text{and thus} \quad \frac{E*(r+\Delta r)}{E*(r)} = \left(\frac{1+\beta}{1-\beta}\right)^{\frac{1}{2\beta}\frac{d}{dr}(\ln f)\Delta r}$$

or

$$\left(\frac{E*(r+\Delta r)}{E*(r)}\right)^{\frac{1}{\Delta r}} = \left(\frac{1+\beta}{1-\beta}\right)^{\frac{1}{2\beta}\left(\frac{d}{dr}(\ln f)\right)}$$

Now $\quad Lim_{\beta\to 0}\left(\frac{1+\beta}{1-\beta}\right)^{\frac{1}{2\beta}} = \exp Lim_{\beta\to 0}\left[\frac{1}{2\beta}\ln\left(\frac{1+\beta}{1-\beta}\right)\right] = e \quad$ and

$$\exp\ln\left(\frac{E*(r+\Delta r)}{E*(r)}\right)^{\frac{1}{\Delta r}} = \exp\frac{1}{\Delta r}[(\ln E*(r+\Delta r) - \ln E*(r)] = \exp\frac{d}{dr}[\ln E*(r)]$$

And hence

$$e^{\frac{d}{dr}[\ln(E*)]} = e^{\frac{d}{dr}[\ln(f)]} \quad \Rightarrow \quad E*(r) = E*(r_0)\frac{f(r)}{f(r_0)}$$

Thus the energy of the boxed photon is influenced by the field in the same manner as is the energy of a massive particle.

Appendix VI. The Einstein Effect and the Shapiro Effect

The Einstein Effect (the deflection of light by a gravitating body)

It turns out that the mathematics governing this effect is the same whether one adopts the usual proper metric, or the T-form. One may write a sort of hybrid form

$$ds^2 = f^{2(c)}c^2 dt^2 - f^{-2k}[dr^2 + f^2 r^2 (d\theta^2 + \sin^2\theta\, d\phi^2)]$$

Recall that we had defined the true value of the speed of light as $c*$ and wrote $c* = f^{(c)}c$ in which c is the proper value, a constant by definition. Note that $(c) = 1$ and $k = 1$ gives the proper metric, while $(c) = 0$ and $k = 2$ gives the T-form.

I introduce this awkward notation in order to make the point that the usual interpretation of the effect holds that half of the deflection is attributable to the curvature of space, and the other half to the reduction of the speed of light: $c* = fc$. According to the Variable Rest Mass interpretation, on the other hand, the whole effect is a consequence of the larger curvature of space, and $c* = c$.

For a light ray in the plane, $\theta = \pi/2$, letting $\phi' = d\phi/dr$,

$$ct = \int_a^\infty f^{-[(c)+k]}(1 + f^2 r^2 \phi'^2)^{1/2}\, dr = \int_a^\infty f^{-2}(1 + f^2 r^2 \phi'^2)^{1/2}\, dr$$

the latter follows since $(c) + k = 2$ for the cases being considered.

Invoking Fermat's principle of least time, produces the appropriate Euler equation, viz.,

$$\partial_{\phi'}[f^{-2}(1 + f^2 r^2 \phi'^2)^{1/2}] = r^2 \phi'(1 + f^2 r^2 \phi'^2)^{-1/2} = \text{constant}$$

Henceforth, we shall write $f_r = (1 - r_S/r)$ and $f_a = (1 - r_S/a)$, in which $r = a$ at perihelion. For $r \to a$, $\phi' \to \infty$, $\Rightarrow$ constant $= a\, f_a^{-1}$. Inserting this gives

$$\phi' = r^{-2}(f_a^2/a^2 - f_r^2/r^2)^{-1/2}$$

Let $\varepsilon = r_S/a$ and $\rho = r/a$. The above becomes

$$d\phi/d\rho = \rho^{-2}[(1 - \varepsilon) - (1 - \varepsilon/\rho)\rho^{-2}]^{-1/2}$$

Expanding to 1st order in ε.

$$d\phi/d\rho = \frac{1}{\rho\sqrt{\rho^2 - 1}}\left[1 + \left(\left(\frac{1}{\rho} + \frac{\rho}{\rho+1}\right)(\varepsilon/2)\right)\right]$$

Now $\displaystyle\int_a^\infty \frac{d\rho}{\rho\sqrt{\rho^2-1}} = -\sin^{-1}(1/\rho)\big|_1^\infty = \pi/2$

Also $\displaystyle\int_a^\infty \frac{d\rho}{\rho^2\sqrt{\rho^2-1}} = \left[\frac{\sqrt{\rho^2-1}}{\rho}\right]_1^\infty = 1$

and $\displaystyle\int_a^\infty \frac{d\rho}{(\rho+1)\sqrt{\rho^2-1}} = \left[\frac{\sqrt{\rho^2-1}}{\rho+1}\right]_1^\infty = 1$

Thus $\phi = \pi/2 + \varepsilon$

The total deflection is $2\varepsilon = 4GM/(c^2 a)$, which agrees with Einstein's famous prediction.

The Shapiro Effect (the delay of radar signals)

We begin with the expression for $d\phi/dr$ deduced above, viz.,

$$\phi' = r^{-2}(f_a^2/a^2 - f_r^2/r^2)^{-1/2}$$

We consider a generalized metric in the central plane, $\theta = \pi/2$

$$dl_k^2 = f_r^{-2k}(dr^2 + f_r^2 r^2 d\phi^2)$$

This includes the proper (k = 1) and T (k = 2) geometries. Then, using the first equation

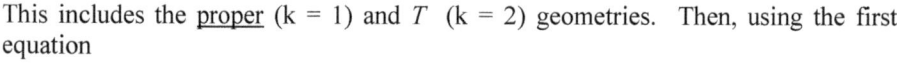

$$dl_k/d\phi = f_r^{-k}[(dr/d\phi)^2 + f_r^2 r^2]^{1/2} = f_r^{-k}[(r^4 f_a^2/a^2 - r^2 f_r^2) + f_r^2 r^2]^{1/2}$$

$$= r^2 f_r^{-k} f_a/a$$

Again, using the first equation we may write

$$dl_k / dr = (dl_k / d\phi)/(dr / d\phi) = (r^2 f_r^{-k} f_a / a) r^{-2} [f_a^2 / a^2 - f_r^2 / r^2]^{-1/2}$$

$$= (r / a) f_a f_r^{-k} [f_a^2 (r / a)^2 - f_r^2]^{-1/2}$$

Again set $c^* = f_r^{(c)} c$. The increment of time world time, dt, is given by

$$dt = dl_k / c^* = f_r^{-(c)} dl_k / c$$

and

$$dt / dr = (1/c^*) dl_k / dr = (1/c)(r/a) f_a f_r^{-[k+(c)]} [f_a^2 (r/a)^2 - f_r^2]^{-1/2}$$

Clearly, the speed of light and the geometry enter in the same way. For brevity, let $p = k + (c)$. Since, $r_S \ll a$ in the case of passage near the limb of the Sun, it will suffice to ignore terms higher than unity in r_S / a. Again let $\varepsilon = r_S / a$ and $\rho = r/a$. Then for the several terms in the expression for dt / dr, we have

$$f_a = (1 - \varepsilon)^{1/2} \approx (1 - \varepsilon / 2)$$

$$f_r^{-p} = (1 - \varepsilon / \rho)^{-p/2} = [1 + (\varepsilon / 2) p / \rho]$$

$$f_a^2 (r/a)^2 - f_r^2 = (1 - \varepsilon)\rho^2 - (1 - \varepsilon / \rho) = \rho^2 - 1 - \varepsilon(\rho^2 - 1/\rho)$$

$$[f_a^2 (r/a)^2 - f_r^2]^{-1/2} = (\rho^2 - 1)^{-1/2} [1 + (\varepsilon / 2)(\rho^3 - 1)/(\rho^3 - \rho)]$$

Inserting these forms, we have, after some reduction,

$$f_a f_r^{-p} [f_a^2 (r/a)^2 - f_r^2]^{-1/2} = (\rho^2 - 1)^{-1/2} \{1 + (\varepsilon / 2)[p(\rho + 1) + 1]/[\rho(\rho + 1)]\}$$

Then, for the time of flight, T, from some location at $r = R$, to the cpa at $r = a$ we have

$$T = \int_a^R (dt / dr) dr = c^{-1} \int_a^R (r/a) f_a f_r^{-p} [f_a^2 (r/a)^2 - f_r^2]^{-1/2} dr$$

$$= (a/c) \int_1^{R/a} \left\{ \frac{\rho}{\sqrt{\rho^2 - 1}} + (\varepsilon / 2) \left[\frac{p}{\sqrt{\rho^2 - 1}} + \frac{1}{(\rho + 1)\sqrt{\rho^2 - 1}} \right] \right\} d\rho$$

Happily, these forms can be integrated in terms of ordinary functions.

$$T = (a/c) \left\{ \sqrt{\rho^2 - 1} + (\varepsilon / 2) \left[p \cosh^{-1} \rho + \sqrt{\frac{\rho - 1}{\rho + 1}} \right] \right\}_1^{R/a}$$

$$T = c^{-1}\sqrt{R^2 - a^2} + (GM/c^3)\left[p \cosh^{-1}(R/a) + \sqrt{\frac{(R/a)-1}{(R/a)+1}} \right]$$

where $p = k + (c)$ Physicists are generally confident that proper geometry is correct, i.e., that $k = 1$, and so are forced to take $(c) = 1$. My choice, of course, is $k = 2$ and $(c) = 0$.

Appendix VII. Regarding Geodesics
While geodesics in relativity are qualitatively different from the geodesics of geometry, one may compare the latter with the relativistic geodesics of light rays.

1. Geometric and relativistic geodesics contrasted

The concept of the geodesic is fundamental in geometry and also in relativity. In ordinary geometry, a geodesic is an extremal, either a local **minimum** (or a maximum) among paths joining two **points** in **three-space**. In relativity, for particles with non-zero rest mass, a geodesic is an extremal, usually a local **maximum** of the time elapsed as recorded by a clock moving with the particle along the geodesic path joining two **events** in **space-time**. This latter circumstance can be understood by taking into account the wave nature of matter, together with Feynman's 'sum over histories' interpretation quantum mechanics. The path 'selected' among all possible paths is that for which 'nearby' paths are in phase with one another, and phase is determined by the 'on-board clock', so to speak. Bear in mind also that the rate of a clock is reduced by the factor $\sqrt{1-(v/c)^2}$, so going too fast is not going to get you there, so to speak.

The on-board clock's time, τ, evolves according to $d\tau = c^{-1}ds$ where

$$ds^2 = g_{jk}dx^j dx^k \quad \tau = c^{-1}\int_A^B \sqrt{g_{jk}\frac{dx^j}{du}\frac{dx^k}{du}}du \quad \text{in which } u \text{ is some parameter.}$$

If the metric is static, as in the case of the Schwarzschild field, we can write

$$ds^2 = g_{00}(dx^0)^2 - g_{\alpha\beta}dx^\alpha dx^\beta \quad \text{where } \alpha \text{ and } \beta \text{ range over (1,2,3).}$$

Now $v^2 = dl_{dt=0}^2 / dt_{fixed}^2 = \dfrac{g_{\alpha\beta}dx^\alpha dx^\beta}{c^{-2}g_{00}(dx^0)^2}$ and $d\tau^2 = c^{-2}g_{00}(dx^0)^2[1-v^2/c^2]$

$$\tau = c^{-1}\int_A^B \sqrt{g_{00}}\sqrt{1-v^2/c^2}\,dt$$

In this form one can see the influence of the gravitational potential $\sqrt{g_{00}}$ and of the motional clock slowing factor $\sqrt{1-v^2/c^2}$.

In contrast, one might say that a photon's clock is frozen, since for a photon, $ds = 0$.

The phase of a photon is directly determined by the time as determined, not by local clocks, but by the 'world time,' which, for an isolated system, is the time kept by 'the clock at infinity.'

This is a consequence of the fact that free photons are not influenced by the force of gravity but move at constant speed along the geodesics of the true geometry (Optical, aka, Telemetric). In the case of photons, the phase is determined by the time where

$$dt^2 = c^{-2}\frac{g_{\alpha\beta}}{g_{00}}dx^\alpha dx^\beta \qquad t = c^{-1}\int_A^B \sqrt{\frac{g_{\alpha\beta}}{g_{00}}\frac{dx^\alpha}{du}\frac{dx^\beta}{du}}\,du$$

Minimizing this time yields the equation for the light path, which is also a geodesic in the true geometry for metrics absent space-time components.

2. Curvature of proper geodesics in the Schwarzschild field

The proper spatial metric in the plane $\theta = \pi/2$ is

$$dl^2 = (1 - r_S/r)^{-1}dr^2 + r^2 d\phi^2$$

$$l = \int_{a-\Delta r}^a \sqrt{(1 - r_S/r)^{-1} + r^2\phi'^2}\,dr$$

The Euler-Lagrange equation is

$$\frac{d}{dr}\frac{\partial L}{\partial \phi'} - \frac{\partial L}{\partial \phi} = 0 \qquad \text{where} \qquad L = \sqrt{(1 - r_S/r)^{-1} + r^2\phi'^2}$$

But $\dfrac{\partial L}{\partial \phi} = 0$ so we have

$$\frac{\partial L}{\partial \phi'} = const = \frac{r^2\phi'}{\sqrt{(1 - r_S/r)^{-1} + r^2\phi'^2}} \quad,$$

Define a as the closest point of approach: $r \to a \implies \phi' \to \infty$,

so $const = a$ and $r^4\phi'^2 = a^2[(1 - r_S/r)^{-1} + r^2\phi'^2]$

$$r^2(r^2 - a^2)\phi'^2 = \frac{a^2}{1 - r_S/r}$$

$$\left(\frac{dr}{d\phi}\right)^2 = \frac{r^2(r^2 - a^2)}{a^2}(1 - r_S/r) \quad \text{Let } \rho = r/a \quad \text{and} \quad \varepsilon = r_S/a$$

$$\left(\frac{d\rho}{d\phi}\right)^2 = (\rho^2 - 1)(\rho^2 - \varepsilon\rho)$$

$$\frac{d\rho}{d\phi} = \sqrt{(\rho^2 - 1)(\rho^2 - \varepsilon\rho)}$$

We shall need $\dfrac{d^2\rho}{d\phi^2} = \dfrac{d}{d\phi}\dfrac{d\rho}{d\phi} = \dfrac{d\rho}{d\phi}\dfrac{d}{d\rho}\dfrac{d\rho}{d\phi} = 1/2\dfrac{d}{d\rho}\left(\dfrac{d\rho}{d\phi}\right)^2$

$$= [4\rho^3 - 3\varepsilon\rho^2 - 2\rho + \varepsilon]/2$$

The radius of curvature is given by

$$R/a = \frac{[\rho^2 + (d\rho/d\phi)^2]^{3/2}}{\rho^2 + 2(d\rho/d\phi)^2 - \rho(d^2\rho/d\phi^2)}$$

Numerator $= [\rho^2 + \rho^4 - \varepsilon\rho^3 - \rho^2 + \varepsilon\rho]^{3/2} = [\rho^4 + \varepsilon\rho(\rho^2 - 1)]^{3/2}$

Denominator $= \rho^2 + 2[\rho^4 - \varepsilon\rho^3 - \rho^2 + \varepsilon\rho] - \dfrac{\rho}{2}[4\rho^3 - 3\varepsilon\rho^2 - 2\rho + \varepsilon]$

$$= \frac{\varepsilon\rho}{2}[3 - \rho^2]$$

The radius of curvature at the cpa is then $(R/a)_{\rho=1} = \dfrac{1}{\varepsilon}$

$$R = \frac{a^2}{2GM/c^2} = \frac{c^2}{2\dfrac{GM}{a^2}} = \frac{1}{2}\frac{c^2}{g}$$

This may be compared with the result for light rays, $R_{light} = \dfrac{1}{3}\dfrac{c^2}{g}$

(see Part II(c), exact solution: photons, p.58)

3. Curvature of telemetric geodesics in the Schwarzschild field

In this case, the spatial metric in the plane $\theta = \pi/2$ is

$$dl^2 = (1 - r_S/r)^{-2} dr^2 + (1 - r_S/r)^{-1} r^2 d\phi^2$$

$$l = \int_{a-\Delta r}^{a} \sqrt{(1-r_S/r)^{-2} + r^2(1-r_S/r)^{-1}\phi'^2}\, dr$$

The Euler-Lagrange equation is

$$\frac{d}{dr}\frac{\partial L}{\partial \phi'} - \frac{\partial L}{\partial \phi} = 0 \quad \text{where} \quad L = \sqrt{(1-r_S/r)^{-2} + r^2(1-r_S/r)^{-1}\phi'^2}$$

But $\dfrac{\partial L}{\partial \phi} = 0$ so we have

$$\frac{\partial L}{\partial \phi'} = const = \frac{r^2(1-r_S/r)^{-1}\phi'}{\sqrt{(1-r_S/r)^{-2} + r^2(1-r_S/r)^{-1}\phi'^2}}$$

Define a as the closest point of approach (cpa): $(r \to a) \Rightarrow (\phi' \to \infty)$,

Thus $const = a\,(1-r_S/a)^{-1/2}$ thus

$$r^4(1-r_S/r)^{-2}\phi'^2 = a^2(1-r_S/a)^{-1}[(1-r_S/r)^{-2} + r^2(1-r_S/r)^{-1}\phi'^2]$$

Let $(1-r_S/r) = u_r$ and $(1-r_S/a) = u_a$

$$[r^4 u_r^{-2} - a^2 r^2 u_a^{-1} u_r^{-1}]\phi'^2 = a^2 u_a^{-1} u_r^{-2}$$

$$r^2[r^2 u_a - a^2 u_r]\phi'^2 = a^2 ; \quad \text{Let } \rho = r/a$$

$$\frac{dr}{d\phi} = \frac{r^2}{a^2}\sqrt{r^2 u_a - a^2 u_r} = r\sqrt{\frac{r^4}{a^4}\left[1-\frac{r_S}{a}\right] - \frac{r}{a}\left[\frac{r}{a} - \frac{r_S}{a}\right]}$$

$$\frac{d\rho}{d\phi} = \rho\sqrt{\rho^4(1-\varepsilon) - \rho(\rho - \varepsilon)}$$

We shall need $\dfrac{d^2\rho}{d\phi^2} = \dfrac{d}{d\phi}\dfrac{d\rho}{d\phi} = \dfrac{d\rho}{d\phi}\dfrac{d}{d\rho}\dfrac{d\rho}{d\phi} = 1/2\dfrac{d}{d\rho}\left(\dfrac{d\rho}{d\phi}\right)^2$

$$= \frac{1}{2}[6(1-\varepsilon)\rho^5 - 4\rho^3 + 3\varepsilon\rho^2]$$

The radius of curvature is given by

$$R/a = \frac{[\rho^2 + (d\rho/d\phi)^2]^{3/2}}{\rho^2 + 2(d\rho/d\phi)^2 - \rho(d^2\rho/d\phi^2)}$$

Numerator $= [\rho^2 + \rho^6(1-\varepsilon) - \rho^4 + \varepsilon\rho^3]^{3/2}$

Denominator $= \rho^2 + 2[\rho^6(1-\varepsilon) - \rho^4 + \varepsilon\rho^2] - \frac{\rho}{2}[6(1-\varepsilon)\rho^5 - 4\rho^3 + 3\varepsilon\rho^2]$

$$= \rho^2 - (1-\varepsilon)\rho^6 + \varepsilon\rho^3/2$$

The radius of curvature at the cpa where $\rho = 1$ is then

$$R_{r=a} = a\frac{[1+(1-\varepsilon)-1+\varepsilon]^{3/2}}{1-(1-\varepsilon)+\varepsilon/2} = \frac{a}{3\varepsilon/2} = \frac{a}{3GM/ac^2} = \frac{1}{3}\frac{c^2}{g}$$

This is identical with the result for the curvature of the path for light rays:

$$R_{light} = \frac{1}{3}\frac{c^2}{g} \qquad \text{(see Part II(c), exact solution: photons, p.58)}$$

110

Appendix VIII. Einstein's Elevator: Motion at constant acceleration *as measured on-board.*

Distances measured by on-board observers are overestimated by the factor $\gamma = 1/\sqrt{1 - v^2/c^2}$, while times are underestimated by the factor γ^{-1}, and thus accelerations are overestimated by the factor γ^3 .

Let $\tilde{a}$ be the acceleration in the 'rest' frame and a be the constant on-board acceleration.

$$\tilde{a} = (1 - v^2/c^2)^{3/2} a$$

If z is the location of a point on the elevator relative to the rest frame, then

$$d^2z/dt^2 = [1 - (dz/dt)^2/c^2]^{3/2} a$$

Introduce $\beta = (dz/dt)/c$. Then

$$(d\beta/dt)/(1 - \beta^2)^{3/2} = a/c$$

$$\beta/\sqrt{1 - \beta^2} = at/c \qquad \text{(We take } \beta = 0 \text{ at } t = 0 \text{)}$$

$$\beta = (at/c)/\sqrt{1 + (at/c)^2}$$

$$dz/dt = c(at/c)/\sqrt{1 + (at/c)^2}$$

$$z = (c^2/a)[\sqrt{1 + (at/c)^2} - 1] + z_0$$

(This result is corroborated by Landau and Lifshitz in *The Classical Theory of Fields, Rev. 2nd Ed.* Pergamon Press, 1962, p.24)

For large t,

$$z - z_0 \approx (c^2/a) \{(at/c)[1 + \frac{1}{2}(c/at)^2 + \cdots] - 1\} \approx ct - c^2/a + \frac{1}{2}\frac{c}{a^2 t} + \cdots$$

Take $z_0 = H$ for the observer 'Hi' at the top of the elevator, and $z_0 = 0$ for the observer 'Lo' at the bottom. Then, if H is the constant (as measured on-board) distance between top and bottom, while a_{Hi} and a_{Lo} are the accelerations measured by Hi and Lo,

$$z_{Hi} - z_{Lo} \approx -c^2 / a_{Hi} + c^2 / a_{Lo} + H$$

But for $t \to \infty$, $\beta \to 1 \Rightarrow (z_{Hi} - z_{Lo}) \to 0$, and thus

$$1 / a_{Hi} = 1 / a_{Lo} + H / c^2$$

It should be possible to show that echo-ranging from Hi to Lo (or vice-versa) will always result in the same value, as measured by the on-board observers. So let's compute the echo-ranging time from ceiling to floor, Hi to Lo, and back.

$$\int_{t_1}^{t_2} (v_{Lo} + c) dt = (z_{Hi} - z_{Lo}) \Big|^{t_1}$$

$$\int_{t_2}^{t_3} (c - v_{Hi}) dt = (z_{Hi} - z_{Lo}) \Big|^{t_2}$$

Consider the first integral

$$c^2 / a_{Lo} \left\{ \sqrt{1 + (a_{Lo} t_2 / c)^2} - \sqrt{1 + (a_{Lo} t_1 / c)^2} \right\} + c(t_2 - t_1)$$

$$= H + c^2 / a_{Hi} \left\{ \sqrt{1 + (a_{Hi} t_1 / c)^2} - 1 \right\} - c^2 / a_{Lo} \left\{ \sqrt{1 + (a_{Lo} t_1 / c)^2} - 1 \right\}$$

$$c^2 / a_{Lo} \sqrt{1 + (a_{Lo} t_2 / c)^2} + c(t_2 - t_1) = H + c^2 / a_{Lo} - c^2 / a_{Hi} + c^2 / a_{Hi} \sqrt{1 + (a_{Hi} t_1 / c)^2}$$

But $H + c^2 / a_{Lo} - c^2 / a_{Hi} = 0$ and we can write

$$c(t_2 - t_1) = (c^2 / a_{Hi}) \sqrt{1 + (a_{Hi} t_1 / c)^2} - (c^2 / a_{Lo}) \sqrt{1 + (a_{Lo} t_2 / c)^2}$$

For the second integral we have similarly

$$c(t_3 - t_2) = (c^2 / a_{Hi}) \sqrt{1 + (a_{Hi} t_3 / c)^2} - (c^2 / a_{Lo}) \sqrt{1 + (a_{Lo} t_2 / c)^2}$$

Subtracting the former from the latter gives

$$c(t_3 + t_1 - 2t_2) = (c^2 / a_{Hi}) \left\{ \sqrt{1 + (a_{Hi} t_3 / c)^2} - \sqrt{1 + (a_{Hi} t_1)^2} \right\}$$

Adding them gives

$$c(t_3 - t_1) = (c^2 / a_{Hi}) \left\{ \sqrt{1 + (a_{Hi} t_3 / c)^2} + \sqrt{1 + (a_{Hi} t_1 / c)^2} \right\} - 2(c^2 / a_{Lo}) \sqrt{1 + (a_{Lo} t_2 / c)^2}$$

Let $\theta_n = a_{Hi}t_n/c$ and $u = a_{Hi}/a_{Lo}$

The above pair of equations may be written as

$$\theta_3 + \theta_1 - 2\theta_2 = \sqrt{1+\theta_3^2} - \sqrt{1+\theta_1^2}$$

$$\theta_3 - \theta_1 = \sqrt{1+\theta_3^2} + \sqrt{1+\theta_1^2} - 2\sqrt{u^2+\theta_2^2}$$

or

$$(\sqrt{1+\theta_1^2} + \theta_1) - (\sqrt{1+\theta_3^2} - \theta_3) = 2\theta_2$$

$$(\sqrt{1+\theta_1^2} + \theta_1) + (\sqrt{1+\theta_3^2} - \theta_3) = 2\sqrt{u^2+\theta_2^2}$$

For brevity let $f_1 = \sqrt{1+\theta_1^2} + \theta_1$, and $f_3 = \sqrt{1+\theta_3^2} - \theta_3$

$$f_1 - f_3 = 2\theta_2$$

$$f_1 + f_3 = 2\sqrt{u^2+\theta_2^2}$$

The squaring the expression for the sum member-by-member and inserting the expression for the difference so as to eliminate θ_2, we have

$$(f_1 + f_3)^2 = 4\{u^2 + \frac{1}{4}(f_1 - f_3)^2\}$$

Whence

$$f_1 f_3 = u^2$$

But $f_3 = \sqrt{1+\theta_3^2} - \theta_3 = \dfrac{1}{\sqrt{1+\theta_3^2} + \theta_3}$ and we can write

$$(\sqrt{1+\theta_1^2} + \theta_1)/(\sqrt{1+\theta_3^2} + \theta_3) = (a_{Hi}/a_{Lo})^2 = (1 - a_{Hi}H/c^2)^2$$

What we wish to prove is that the time as measured on-board – say by Hi – for echo-ranging off the opposite end of the elevator is constant throughout the motion. Hi's clock will run slow by the factor $\sqrt{1-\beta_{Hi}^2}=1/\sqrt{1+(a_{Hi}t/c)^2}$. Thus letting τ represent Hi's time

$$\tau_3-\tau_1=\int_{t_1}^{t_3}\frac{dt}{\sqrt{1+(a_{Hi}t/c)^2}}=(c/a_{Hi})\int_{\theta_1}^{\theta_3}\frac{d\theta}{\sqrt{1+\theta^2}}=(c/a_{Hi})[\sinh^{-1}\theta_3-\sinh^{-1}\theta_1]$$

$$=(c/a_{Hi})\{\ln(\sqrt{1+\theta_3^2}+\theta_3)-\ln(\sqrt{1+\theta_1^2}+\theta_1)\}$$

$$=(c/a_{Hi})\ln\left[\frac{\sqrt{1+\theta_3^2}+\theta_3}{\sqrt{1+\theta_1^2}+\theta_1}\right]=(c/a_{Hi})\ln(1-a_{Hi}H/c^2)^{-2}$$

which is a constant, as was to be proved.

Incidentally, using $\ln(1-x)^{-1}\approx x+x^2+x^3+\cdots$ the time of flight is approximately

$$\tau_3-\tau_1\approx 2(c/a_{Hi})[a_{Hi}H/c^2+(a_{Hi}H/c^2)^2+\cdots]=2H/c[1+a_{Hi}H/c^2+\cdots]$$

Thus Einstein's assumption that the measured speed of light would be unaffected in the accelerating elevator is invalid, and the equivalence principle fails to hold, since in a gravitational field, the measured speed is in fact unaffected.

Appendix IX. An Intuitive derivation of the Kerr flow velocity

In the abstract treatment of general stationary fields, the expression $\beta^2 = 1 - g_{00} g^{00}$ was deduced. Here I present the derivation for the particular case of the Kerr field.

First we find the form taken by the above equation for β^2 in the case of a Kerr-type metric, in which only $g_{0\phi}$ space-time components occur.

$$g_{ij} = \begin{bmatrix} g_{00} & g_{0\phi} & 0 & 0 \\ g_{0\phi} & g_{\phi\phi} & 0 & 0 \\ 0 & 0 & g_{rr} & 0 \\ 0 & 0 & 0 & g_{\theta\theta} \end{bmatrix} \qquad g^{ij} = \begin{bmatrix} g_{\phi\phi}/d & -g_{0\phi}/d & 0 & 0 \\ -g_{0\phi}/d & g_{00}/d & 0 & 0 \\ 0 & 0 & g_{rr}^{-1} & 0 \\ 0 & 0 & 0 & g_{\theta\theta}^{-1} \end{bmatrix}$$

where $d = g_{00} g_{\phi\phi} - g_{0\phi}^2$

$$\beta^2 = 1 - g_{00} g^{00} = 1 - g_{00} g_{\phi\phi}/d = -\frac{g_{0\phi}^2}{d} = \frac{-g_{0\phi}^2}{g_{00}g_{\phi\phi} - g_{0\phi}^2} = \frac{g_{0\phi}^2}{g_{00}(-g_{\phi\phi}) + g_{0\phi}^2}$$

Note that $g_{\phi\phi} < 0$ if one takes, as I have always done, $g_{00} > 0$. Then

$$\bullet \qquad \beta = \frac{g_{0\phi}/\sqrt{g_{00}(-g_{\phi\phi})}}{\sqrt{1 + \left[g_{0\phi}/\sqrt{g_{00}(-g_{\phi\phi})}\right]^2}}$$

It is also the case that in the abstract treatment of general stationary fields, I relied on the formal definition of the pure space metric, γ_{ij}, as defined by Landau & Lifshitz. This metric is constructed via an echo ranging ansatz, using Einstein's assumption of equal times: $t_{AB} = t_{BA}$, which is incorrect in general, and is incorrect in the case of the Kerr solution. (It turns out that SR effects compensate, and the deduced expression for velocities is correct in spite of γ_{ij} not representing the true geometry.) To simplify notation introduce the representation

$$ds^2 = g_{00}\left(dx^o\right)^2 - 2\bar{g}_{0\phi}\, dx^o d\phi - \bar{g}_{\phi\phi}\, d\phi^2$$

where $\bar{g}_{\phi\phi} > 0$, and the other, non–involved components are ignored.

Constraining a light pulse to a circle, we set $ds = 0$ and solve for dx^0

$$dx^0 = \sqrt{\frac{\bar{g}_{\phi\phi}}{g_{00}}}\left[\sqrt{1+\left(\frac{\bar{g}_{0\phi}}{\sqrt{g_{00}\bar{g}_{\phi\phi}}}\right)^2} \pm \left(\frac{\bar{g}_{0\phi}}{\sqrt{g_{00}\bar{g}_{\phi\phi}}}\right)\right] \equiv \sqrt{\frac{\bar{g}_{\phi\phi}}{g_{00}}}\left[\sqrt{1+\Phi^2} \pm \Phi\right]$$

$$\Delta x^0 = \int_0^{2\pi}\sqrt{\frac{\bar{g}_{\phi\phi}}{g_{00}}}\left[\sqrt{1+\Phi^2} \pm \Phi\right] = 2\pi\sqrt{\frac{\bar{g}_{\phi\phi}}{g_{00}}}\left[\sqrt{1+\Phi^2} \pm \Phi\right]$$

But if $2\pi r *$ is the telemetric circumference, then $\Delta x^0 = c\dfrac{2\pi r *}{c \mp v} = \dfrac{2\pi r *}{1 \mp \beta}$,

whence

$$\left[\sqrt{\frac{g_{00}}{\bar{g}_{\phi\phi}}}r *\right]\frac{1}{1\mp\beta} = \sqrt{1+\Phi^2} \pm \Phi \text{ or, setting } R = \sqrt{\frac{g_{00}}{\bar{g}_{\phi\phi}}}r *,$$

$$\frac{R}{1\mp\beta} = \sqrt{1+\Phi^2} \pm \Phi$$

Dividing the expression with the lower sign by that of the upper sign gives

$$\frac{1-\beta}{1+\beta} = \frac{\sqrt{1+\Phi^2}-\Phi}{\sqrt{1+\Phi^2}+\Phi} = \frac{1-\Phi/\sqrt{1+\Phi^2}}{1+\Phi/\sqrt{1+\Phi^2}} \quad \text{whence}$$

- $\beta = \dfrac{\Phi}{\sqrt{1+\Phi^2}}$, which is identical to previous result, since $\Phi = \dfrac{\bar{g}_{0\phi}}{\sqrt{g_{00}\bar{g}_{\phi\phi}}}$

We may also derive an expression for $r *$, the telemetric circumference $/2\pi$.

Adding the two expressions in $\dfrac{R}{1\mp\beta} = \sqrt{1+\Phi^2} \pm \Phi$ gives

$$\frac{R}{1-\beta^2} = \sqrt{1+\Phi^2} \text{ but } 1-\beta^2 = \frac{1}{1+\Phi^2} \quad \text{so}$$

$$R = \frac{1}{\sqrt{1+\Phi^2}} \text{ and finally, } r * = \sqrt{\frac{\bar{g}_{\phi\phi}}{g_{00}}}R = \frac{\bar{g}_{\phi\phi}}{\sqrt{g_{00}\bar{g}_{\phi\phi}+\bar{g}_{0\phi}^2}}$$

Appendix X.   A procedure for finding the stenosurface of the Kerr solution for any value of the polar angle.

In Part II.(c), the following equation for the stenosurface was deduced:

$$2r^7 - 3r_S r^6 + 2(1 + 2C^2)a^2 r^5 - (3 + 2C^2)r_S a^2 r^4$$
$$+ [(4C^2 + 2C^4)a^2 + 3S^2 r_S^2]a^2 r^3 - (C^4 + 2S^4)r_S a^4 r^2$$
$$+ (2C^2 a^2 - S^2 r_S^2)C^2 a^4 r + (1 + S^2)C^2 r_S a^6 = 0$$

where $C = \cos\theta$ and $S = \sin\theta$. Introduce $\rho = r/r_S$ and $\alpha = a/r_S$

$$2\rho^7 - 3\rho^6 + 2(1 + 2C^2)\alpha^2 \rho^5 - (3 + 2C^2)\alpha^2 \rho^4$$
$$+ [(4C^2 + 2C^4)\alpha^2 + 3S^2]\alpha^2 \rho^3 - (C^4 + 2S^4)\alpha^4 \rho^2$$
$$+ (2C^2 \alpha^2 - S^2)C^2 \alpha^4 \rho + (1 + S^2)C^2 \alpha^6 = 0$$

After dividing the last equation by ρ^6 introduce $A = (\alpha/\rho)^2$

$$2\rho - 3 + 2(1 + 2C^2)A\rho - (3 + 2C^2)A$$
$$+ (4C^2 + 2C^4)A^2 \rho + 3S^2 A\rho^{-1} - (C^4 + 2S^4)A^2$$
$$+ 2C^4 A^3 \rho - S^2 C^2 A^2 \rho^{-1} + (1 + S^2)C^2 A^3 = 0$$

Multiplying by ρ produces a quadratic equation in ρ :

$$2\,[1 + (1 + 2C^2)A + (2C^2 + C^4)A^2 + C^4 A^3]\rho^2$$
$$- [3 + (3 + 2C^2)A + (C^4 + 2S^4)A^2 - (1 + S^2)\,C^2 A^3]\rho$$
$$+ [3 - C^2 A]S^2 A = 0$$

In this indirect procedure, a value for A is assumed and the above is 'solved' for ρ, after which the value for α is found from the definition of A : $\alpha = \rho\sqrt{A}$.

Appendix XI. The Massless Kerr Field is flat but interesting.

The Kerr Metric in the usual coordinates can be written

$$ds^2 = \rho^{-2}(\rho^2 - r_S r)(dx^0)^2 + 2\rho^{-2}r_S ra \sin^2\theta \, d\phi \, dx^0$$
$$- \rho^{-2}u^4 \sin^2\theta \, d\phi^2 - \Delta^{-1}\rho^2 \, dr^2 - \rho^2 \, d\theta^2$$

Here we have introduced the expressions

$$r_S = 2GM/c^2 \text{ and } a = L/Mc,$$ where L and M are the angular momentum and mass of the rotating object, respectively, as well as the expressions

$$\rho^2 = r^2 + a^2\cos^2\theta, \quad \Delta = r^2 + a^2 - r_S r \quad \text{and}$$

$$u^4 = \rho^2(r^2 + a^2) + r_S r a^2 \sin^2\theta$$

Setting $M = 0 \Rightarrow r_S = 0, \quad \Delta = (r^2 + a^2), \quad \rho^{-2}u^4 = (r^2 + a^2),$ whence

$$ds^2 = (dx^0)^2 - \rho^{-2}u^4\sin^2\theta \, d\phi^2 - \Delta^{-1}\rho^2 \, dr^2 - \rho^2 \, d\theta^2 \equiv (dx^0)^2 - dl^2$$

$$dl^2 = \frac{r^2 + a^2\cos^2\theta}{r^2 + a^2} dr^2 + (r^2 + a^2\cos^2\theta)d\theta^2 + (r^2 + a^2)\sin^2\theta \, d\phi^2$$

The parameter a has been identified as a length which is proportional to the angular momentum per unit mass of a presumed rotating body, but it can just as well be considered to be a mere constant of integration. The result of setting $M = 0$ and $a \neq 0$ still represents a valid solution of the equations for a gravitational field, even though, since $g_{00} = 1$, there is no gravitational field as such. Also, perhaps surprisingly, there is no rotation. Furthermore, as we are about to prove, the space defined by the above metric is flat. Nevertheless, it turns out that it is not identical with Euclidean space, but rather seems to incorporate a magical ring-singularity connecting two flat spaces, each of which is Euclidean, except on the ring, of course.

Define new variables as follows

$$Z = r\cos\theta, \quad R = \sqrt{r^2 + a^2}\sin\theta, \quad \phi = \phi$$

$$dZ = \cos\theta \, dr - r\sin\theta \, d\theta, \quad dR = \frac{r\sin\theta}{\sqrt{r^2 + a^2}} dr + \sqrt{r^2 + a^2}\cos\theta \, d\theta, \quad d\phi = d\phi$$

$$dZ^2 = \cos^2\theta \, dr^2 - 2r\cos\theta\sin\theta \, dr \, d\theta + r^2\sin^2\theta \, d\theta^2$$

$$dR^2 = \frac{r^2\sin^2\theta}{r^2 + a^2} dr^2 + 2r\sin\theta\cos\theta \, dr \, d\theta + (r^2 + a^2)\cos^2\theta \, d\theta^2$$

$$dR^2 + dZ^2 = \frac{(r^2 + a^2)\cos^2\theta + r^2\sin^2\theta}{r^2 + a^2}dr^2 + [(r^2 + a^2)\cos^2\theta + r^2\sin^2\theta]d\theta^2$$

$$= \frac{r^2 + a^2\cos^2\theta}{r^2 + a^2}dr^2 + (r^2 + a^2\cos^2)d\theta^2$$

Finally, then, it has been proven that the space is flat, since a simple coordinate transformation brings the metric to the representation of flat space in cylindrical coordinates:

$$dR^2 + dZ^2 + R^2 d\phi^2 = \frac{r^2 + a^2\cos^2\theta}{r^2 + a^2}dr^2 + (r^2 + a^2\cos^2\theta)d\theta^2 + (r^2 + a^2)\sin^2\theta\, d\phi^2$$

But the original space is not identical with the R, Z, ϕ space, since $R \geq a$ unless $\sin\theta = 0$.

It is useful to plot curves of Z versus R with r as a parameter, on the one hand, and Z versus R with θ as a parameter. From the definitions

$$Z = r\cos\theta, \quad R = \sqrt{r^2 + a^2}\sin\theta$$

One has $\quad \dfrac{R^2}{r^2 + a^2} + \dfrac{Z^2}{r^2} = 1 \quad$ and $\quad \dfrac{R^2}{\sin^2\theta} = \dfrac{Z^2}{\cos^2\theta} + a^2 \quad$ or

$$\frac{(R/a)^2}{(r/a)^2 + 1} + \frac{(Z/a)^2}{(r/a)^2} = 1 \quad \text{and} \quad \frac{(R/a)^2}{\sin^2\theta} - \frac{(Z/a)^2}{\cos^2\theta} = 1$$

These are confocal elliptic and hyperbolic families of curves which share common foci.

This figure is to be thought of as a distortion of spherical coordinates r, θ, ϕ in which the origin has been opened up into a disk (seen edge-on in the figure) so that the lines of constant polar angle are no longer straight lines meeting at a point, but rather hyperbolae intercepting the disk at a point lying at the distance $a\sin\theta$ from the center of the disc. For large values of r, the figure approaches that of the standard spherical coordinates.

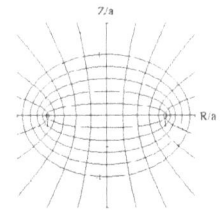

The important thing is that if the parameter r is positive in the upper half of the figure, and if it is to vary smoothly through the disc, it will be negative in the lower half of the figure. This is the 'new' region which has been connected to the 'normal' region in which r is positive. If one imagines that a traveller passes through the ring from above so as to enter the new region, he can return to the normal region by either reversing his track, or by following a path that encircles the ring so as to pass through the disk again from above. It's rather like 3-D version of a Riemann surface of two sheets.

Appendix XII. A Self-consistent Derivation of Rest Mass Reduction

We assume that when a body is raised in a gravitational field, its mechanical energy is increased, and that this increase in energy will result in an increase in its rest mass in accordance with the equation, $E = mc^2$. Specifically, we find $m* = m\sqrt{1 - r_S/r}$.

Lemma: We first show that for a body falling from rest at r_0 in the Schwarzschild field, the telemetric acceleration, $g*$ has the initial value, $g*(r_0) = -GM/r_0^2$. The integral of motion in this case is

$$\sqrt{(1 - r_S/r}/\sqrt{1 - v^2/c^2} = \sqrt{1 - r_S/r_0} \text{ , whence,}$$

$$v = c\sqrt{(r_S/r - r_S/r_0)}/\sqrt{1 - r_S/r_0}$$

But assuming the VRM metric, $v = v* = dR*/dt = (1 - r_S/r)^{-1} dr/dt$ whence

$$dr/dt = (1 - r_S/r)v = c(1 - r_S/r)\sqrt{(r_S/r - r_S/r_0}/\sqrt{1 - r_S/r_0}$$

Now $v* = v$, and $g* = dv*/dt$, thus,

$$g* = dv/dt = -\left(\tfrac{1}{2}cr_S/r^2\right)/(\sqrt{r_S/r - r_S/r_0}\sqrt{1 - r_S/r_0} \text{) } dr/dt$$

Inserting the expression for dr/dt into the above equation for $g*$ yields

$$g* = -(\tfrac{1}{2}cr_S/r^2)(1 - r_S/r)/(1 - r_S/r_0), \text{ and finally,}$$

$$g*(r_0) = -\tfrac{1}{2}c^2r_S/r_0^2 = -GM/r_0^2$$

Dropping the subscript, $_0$, the '$g*$ force' acting on a body at rest at r, is

$$g* = GM/r^2 \text{ , and thus we can write}$$

$$dm*c^2 = m*g*dR* = (GMm*/r^2)dR*, \text{ but } dR* = dr/(1 - r_S/r) \text{ so}$$

$$dm*/m* = (GM/c^2)/r^2(dr/(1 - r_S/r)) = (1/2)r_S/r^2(dr/(1 - r_S/r))$$

$$dm*/m* = d\sqrt{1 - r_S/r}/\sqrt{1 - r_S/r} \text{ , whence,}$$

$$m* = m\sqrt{1 - r_S/r}$$

Admittedly, the above reasoning is circular in that the validity of the VRM metric is at the outset presumed to be correct. Nevertheless, it is satisfying that the exact expression for the dependence of rest mass results from the idea that the gravitational potential energy of a body resides in the body itself.

Appendix XIII. The Gaffes of Relativity

1. The greatest Gaffe: The relativity of time.

Here I quote Kip Thorne: [59]

> "Within days after formulating his equivalence principle, Einstein used it to make an amazing prediction, called *gravitational time dilation: If one is at rest relative to a gravitating body, then the nearer one is to the body, the more slowly one's time must flow.*" [Emphasis Thorne's]

In his 1911 paper, Einstein deduced that clock rates must be reduced in a gravitational field. He offered no explanation of this effect, and did not refer to the 'flow of time.'

What causes the slowing of clocks? Modern day physicists do address this issue, but in the most superficial manner imaginable: they claim that the 'flow of time' is reduced in a gravitational field, and that clocks are slowed because they are driven, so to speak, by the 'flow of time!' Einstein's phrase 'clock slowing' is replaced by the phrase 'gravitational time dilation.' This hardly qualifies as a scientific inquiry into the cause of clock slowing: – it is just a vague, unwarranted, *ad hoc* assumption, leading nowhere.

According to this view, light also moves in response to the 'flow of time,' so that local observers will find that the distance traveled divided by the reduced time will equal the constant, c. Everything is in slow motion – the progress of light as well as the hands of the clock. But the distance traveled divided by the corresponding interval of true time (measured by the clock at infinity) will be smaller than c. In this sense, the speed of light could be said to be reduced, just as Einstein mistakenly assumed in his 1911 paper.

In his discussion of the gravitational red shift, Thorne repeatedly asserts that the rate at which a clock runs is determined by the rate at which 'time flows' at the location of the clock.

Even Feynman seemed not to object to this concept of time:[60]

> "One way of describing this situation is to say that the time scale is faster at the top; *time flows are different in different gravitational potentials...*" [Emphasis added]

---

[59] Kip S. Thorne (1994) *Black Holes & Time Warps: Einstein's Outrageous Legacy.* W. W. Norton & Company, New York, NY, p.100.

[60] Richard P. Feynman, Fernando B. Morinigo, and William G. Wagner (1995) *Feynman Lectures on Gravitation*, Addison-Wesley Publishing Company, Reading MA, p.94.

Einstein, to the best of my knowledge, never referred to the 'flow of time.' Einstein recognized that the gravitational red shift was properly understood to be a direct consequence of the slowing of clock rates. Also, the phrase *gravitational time dilation* does not appear in any of his discussions of this phenomenon.

In his 1911 paper, Einstein says

> "Nothing compels us to assume that [identical] clocks …in different gravitational potentials must be regarded as going at the same rate."

Again, in his 1916 presentation of the general theory, Einstein writes

> "Thus the clock goes more slowly if set up in the neighbourhood of ponderable masses."

Finally, in his popularization, *Relativity*, Einstein declares

> "…we can regard an atom which is emitting spectral lines as a clock, so that the following statement will hold: ***An atom absorbs or emits light of a frequency which is dependent on the potential of the gravitational field in which it is situated.***" [Einstein's emphasis]

The amazing fact is that neither Einstein nor any of his followers ever conceived that it was a worthwhile endeavor to try to discover the **cause** of gravitational clock slowing (frequency reduction). Perhaps the air of mystery which enshrouds all things relativistic precluded such a naïve endeavor.

But is this concept of variable time flows really a viable hypothesis?

In order that the locally determined speed of light be invariant, the interpretation invoking the idea of a reduced 'flow of time' requires that local measuring rods be unaffected, and thus the interpretation holds that proper measurements correctly describe the geometry of a black hole. But the Abramowicz effect – the reversal of the direction of centrifugal force inside the orbit of photons – is incompatible with proper geometry, as has been demonstrated. This is a valid argument against the variable flow of time concept, rendering it untenable.

Regarding the integral of motion for a test mass falling freely in the Schwarzschild field, viz.,

$$\frac{\sqrt{1-r_s/r}\ mc^2}{\sqrt{1-v^2/c^2}} = \frac{m^*c^2}{\sqrt{1-v^2/c^2}} = const,$$

the variable rest mass interpretation clearly identifies this as a statement of energy conservation; the variable time-flow concept does not.

At a deeper level, it may be said that accepting the concept of variable time flows is analogous to accepting intelligent design as an 'explanation' for biological complexity: acceptance precludes any possibility of further analysis.

In a closely related matter, neither Einstein nor any of his followers (excepting Julian Schwinger) took note of the fact that Einstein's analysis of the red shift proves not only that the reduction occurs at the source, but also that the emitted light moves with constant frequency between source and detector. Clearly, this means that photons and other zero rest mass particles, including hypothetical gravitons, are not subject to the force of gravity. Nor do they act as sources. This fact is of inestimable importance, but again, no one seems to realize that Einstein's argument of 1911 proved the truth of it.

Kip Thorne's Time Machine
Thorne's peculiar assumption regarding the 'flow of time' is central to arguments he presents in a speculation regarding the use of 'wormholes' as elements of a machine enabling time travel.

In Thorne's view, the 'flow of time' determines the rate of a local clock and thus the true 'time' corresponds exactly to the indication of the clock. In contrast, the analysis presented in this book establishes the cause of the slowing of clocks. The special relativity effect is shown to be a direct consequence of the wave nature of matter, and the gravitational effect is shown to result from the reduction of rest mass. In either case, the clocks fail to keep the true time, which corresponds to the time kept by an ideal clock far removed from gravitating masses and at rest in the aether (the reality of which was established in the consideration of the flat three-torus).

In his simplest version of a time machine, Thorne arranges to cause one of the two 'mouths' of a wormhole to move at highly relativistic speed on a ten year round trip, as assessed by *external* observers, while the other mouth remains at rest. But this special wormhole is able to maintain a constant *internal* length, mouth to mouth, of just three meters. A clock just outside the travelling mouth will be dramatically slowed, so that on return from its ten year sojourn, it will have advanced by just twelve hours. But the clock just inside the same mouth, having not moved at all relative to the stationary mouth, will have advanced the full ten years. From the point of view presented in this book (reluctantly accepting the possibility  of this imaginary scenario), one merely notes a discrepancy in the reading of the two clocks near the mouth that moved: full stop. From this follows no conclusion regarding time travel. But Thorne, with his curious concept of the 'flow of time' believes that the 'time' just outside the mouth that moved is just twelve hours later than the time of launch. Just inside that mouth, he claims, the 'time' is ten years after the launch time. So if one traverses the wormhole from stationary end to moving end, one moves back in time by 10 years, and, conversely, taking the tube in the opposite direction takes one 10 years into the future.

2. Time travel nonsense

Here we are considering returning to the past: everyone, like it or not, is impelled into the future at a rate which is quite out of one's control.

The past, it seems to me, is a *definite* thing: a definite sequence of events ordered in a definite way in space and time. If this definition of the past is accepted, then if one returns to the past, one must do so in a way that leaves intact every facet of historical fact. This stipulation includes the exact sequence of mental states of all persons alive at the time, including any who might be intruders from the future.

For such intruders, there can be no memory of their future unless such 'memories' had been experienced as premonitions. Furthermore, the time traveler is constrained to do precisely what he had done: he cannot choose, or even think, to do otherwise. If these constraints were to be violated, the traveler would not be in the past, but in some ersatz pseudo-past.

The traveler thus has no free will, but he cannot be aware of this absolute constraint on every aspect of his behavior. Of course, it is manifestly obvious that he can only return to times and places in which he actually lived.

A little reflection shows that within this definition of the past, any and all of us may be time travelers from the future: there is no way to tell. Within this the uniquely reasonable definition of the past, the whole concept of travel to the past becomes vacuous, absurd.

It is clear that when Kip Thorne and other experts in relativity talk about returning to the past in time machines, they are really talking about visiting an ersatz-past: a 'parallel universe,' the history of which is as much like the real past as possible, but just enough different so as to accommodate the intruder from the future. It is not clear to me whether these experts imagine that such parallel universes are created on the spot to accommodate the junketing time traveler, or whether they have an (almost) independent existence.

3. The bowling ball on the rubber sheet mal-analogy

This attempted analogy suggests that the curvature of three-space gives rise to gravitational acceleration. This is quite untrue: in fact, the effect is entirely determined by what might be called the 'curvature of time,' that is, by the spatial dependence of the time-time component, g_{00}, of the metric tensor.

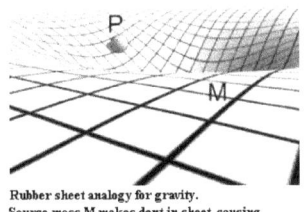

Rubber sheet analogy for gravity.
Source mass M makes dent in sheet, causing target body at P to roll "downhill" toward M.

Consider a time-independent field in which all time-space components of the metric tensor vanish (no aether flows). In this case the invariant interval may be written

$$ds^2 = g_{00}(dx^o)^2 - g_{\alpha\beta}dx^\alpha dx^\beta$$

in which $g_{\alpha\beta}$ is the three space metric.

The Christoffel symbols are generally,

$$\Gamma^i_{kl} = 1/2 \; g^{im}\left[\frac{\partial g_{km}}{\partial x^l} + \frac{\partial g_{lm}}{\partial x^k} - \frac{\partial g_{kl}}{\partial x^m}\right] \qquad \text{(Latin indices 0,1,2,3; Greek 1,2,3)}$$

In particular,

$$\Gamma^0_{kl} = 1/2 \; g^{00}\left[\frac{\partial g_{k0}}{\partial x^l} + \frac{\partial g_{l0}}{\partial x^k} - \frac{\partial g_{kl}}{\partial x^0}\right] = 1/2 \; g^{00}\left[\frac{\partial g_{k0}}{\partial x^l} + \frac{\partial g_{l0}}{\partial x^k}\right]$$

Now one of k *and* l must be zero (0), and the other must be among 1, 2, 3. Thus the only non-zero symbols are

$$\Gamma^0_{0\alpha} = 1/2 \; g^{00}\frac{\partial g_{00}}{\partial x^\alpha}. \quad \text{Also,} \quad \Gamma^0_{00} = 0.$$

The equations of motion. First consider the 'motion' of x^0

$$\frac{d^2x^0}{ds^2} = \Gamma^0_{0\alpha}\frac{dx^0}{ds}\frac{dx^\alpha}{ds} = \frac{dx^0}{ds}1/2 \; g^{00}\frac{\partial g_{00}}{\partial x^\alpha}\frac{dx^\alpha}{ds}$$

Clearly, if g_{00} is not a function of position, or time, then $\dfrac{d^2x^0}{ds^2} = 0$, and

$$ds = (const)dx^0.$$

That is, the proper time interval for a clock possibly in motion is some constant times the world time interval. Let $d\tau_{fix}$ represent a time interval for an unmoving (fixed) clock:

$(d\tau_{fix})^2 = c^{-2}ds^2\big|_{dx^\alpha=0} = c^{-2}g_{00}(dx^0)^2$. While for a clock possibly in motion,

$d\tau^2 = c^{-2}ds^2 = c^{-2}g_{00}(dx^0)^2 - c^{-2}g_{\alpha\beta}dx^\alpha dx^\beta$. Then

$(d\tau/d\tau_{fix})^2 = 1 - c^{-2}\dfrac{g_{\alpha\beta}dx^\alpha dx^\beta}{(d\tau_{fix})^2} = 1 - c^{-2}\dfrac{dl^2}{(d\tau_{fix})^2} = 1 - (v/c)^2$

Now $ds/dx^0 = \sqrt{g_{00}}(d\tau/d\tau_{fix})$, so $ds = \sqrt{g_{00}}\sqrt{1-(v/c)^2}\,dx^0$, and we have identified the (*const*) above, and also demonstrated that a test particle moves with constant speed when g_{00} is not a function of position.

In the other equations of motion, $\dfrac{d^2x^\alpha}{ds^2} = \Gamma^\alpha_{kl}\dfrac{dx^k}{ds}\dfrac{dx^l}{ds}$, we consider the case where just one of k or l is zero.

Now, $\Gamma^\alpha_{0k} = 1/2\,g^{\alpha m}\left[\dfrac{\partial g_{0m}}{\partial x^k} + \dfrac{\partial g_{km}}{\partial x^0} - \dfrac{\partial g_{0k}}{\partial x^m}\right]$.

Since $g^{\alpha 0} = 0$, $m \to \mu \Rightarrow g_{0\mu} = 0$.

Thus $\Gamma^\alpha_{0k} = 1/2\,g^{\alpha\mu}\left[-\dfrac{\partial g_{0k}}{\partial x^\mu}\right]$,

But by the no-flow assumption, $g_{0\alpha} = 0$. Thus the only Christoffel symbol with index zero (time) is

$\Gamma^\alpha_{00} = 1/2\,g^{\alpha\mu}\left[-\dfrac{\partial g_{00}}{\partial x^\mu}\right]$

Again, if g_{00} does not depend upon position, this Christoffel symbol, representing the gravitational force, vanishes. The remaining equations of motion,

$\dfrac{d^2x^\alpha}{ds^2} = \Gamma^\alpha_{\lambda\mu}\dfrac{dx^\lambda}{ds}\dfrac{dx^\mu}{ds}$,

are just the equations defining the geodesics of the three space.

Thus *if g_{00} is a constant, test particles move along the geodesics of three-space at constant speed.* Also, regardless of how curved the three space might be, there is no gravitational force unless g_{00} varies with position.

Incidentally, another error is common in such pictures (as in the lower figure above): The surface of the sheet as depicted quickly becomes quite flat, which would correspond to a short range field. But the field of gravity is long range, demanding that the sheet rise parabolically with distance. Finally, modern writers often refer to 'spacetime warping.' I believe that this inappropriate description arises from a misunderstanding of the embedding diagram, which does have the appearance of a warped surface, but which really just informs us that there is an excess of space near a gravitating body – no 'warpage' at all.

130

4. Red Shift Nonsense

Misner, Thorne and Wheeler, [61] and Bernard Schutz [62] present identical arguments to

'explain' the gravitational red shift. A particle of mass m is dropped from a tower of height h. It reaches the ground with energy $mc^2 + mgh$, all of which is converted into a photon which is sent back up the tower to become a particle again, now of mass m'. It is claimed that the photon must lose energy, so that $m' = m$; otherwise perpetual motion would be possible. The error is in the unconscious assumption that the initial energy of the particle was equal to mc^2, whereas, since it was at the top of the tower, its total energy was $mc^2 + mgh$, so that its *true* rest mass (as opposed to its proper rest mass) was indeed equal to m'. MT&W explain: "The energy of the photon must decrease just as that of a particle does when it climbs out of the gravitational field."

Schutz agrees: "…a photon climbing in the Earth's gravitational field will lose energy (not surprisingly) and will consequently be redshifted." [Schutz's parenthetical expression].

Robert Wald [63] goes along with this idea, remarking, "Physically, this makes sense…we would expect that the photon energy to be degraded as it 'climbs out of the gravitational well.'"

[61] Charles W Misner, Kip S Thorne and John A Wheeler (1970) *Gravitation* W H Freeman & Co., p. 187

[62] Bernard F Schutz, (1985) *A first course in general relativity* Cambridge University Press, pp. 119-120

[63] Robert M Wald (1984) *General Relativity* University of Chicago Press, p. 137

Finally, Schwinger [64] gets it right. Having derived the effect, he remarks:

"Note that we have taken for granted that the frequency of the light emitted by the source is received by the detector as light of the *same* frequency. One sometimes encounters a different explanation of the gravitational red shift. A photon rising up against a gravitational attraction loses energy and therefore suffers a decrease in frequency. But the energy of a photon, like that of any other projectile (friction aside) is *conserved*; its frequency does *not* change. Rather, it is the *standards* of frequency that differ at different locations." [Emphasis Schwinger's]

5. Lorentz Invariance misapplied

Lorentz invariance pertains to the equations of physics. It is not a fact of nature that applies to everything. In particular, Lorentz invariance is not to be imposed upon the contents of the universe. Yet modern day physicists insist that all observers, regardless of their motion, will not differ in their perception of the universe. In particular, the electromagnetic field, they insist, will appear to be homogeneous and isotropic to all observers. The fact that the cosmic microwave background radiation shows an obvious dipole structure, blue-shifted in one direction, red-shifted opposite has not deterred the theorists – they still insist on the general validity of 'boost symmetry' as applied to other fields. Importantly, in dealing with the cosmical constant they invoke boost symmetry, which in this case, demands that ρ, the effective mass density of space, if taken to be positive, must be accompanied by a *negative* pressure: $p = -\rho c^2$. This implies that space itself is somehow in tension, a difficult concept, to put it mildly.

---

[64] Julian Schwinger (1986) *Einstein's Legacy* Scientific Books Inc. p. 142

Appendix XIV   Correction to Standing Wave Model for Extended Particles, accounting for unequal amplitudes in the standing wave.

Consider photons reflected off the advancing 'mirror.' Just reflected photon A moves distance d in time d/c. Approaching Photon B, located distance d from the 'mirror' at time zero, will hit the plate after time t such that

$$vt + ct = d , \quad \text{so} \quad t = \frac{d}{c}\frac{1}{1+\beta}$$

Thus at time t photon B will be located at $vt = \frac{\beta}{1+\beta}d$

The distance between A and B at time t is thus reduced from d to $\frac{1-\beta}{1+\beta}d$ and this

implies that the photon number density satisfies

$$n_\rightarrow = \frac{1+\beta}{1-\beta}n_\leftarrow$$

The total number density is $\quad n = n_\rightarrow + n_\leftarrow = \frac{2}{1+\beta}n_\rightarrow = \frac{2}{1-\beta}n_\leftarrow \quad$ so

$$n_\rightarrow = \frac{n}{2}(1+\beta) \quad \text{and} \quad n_\leftarrow = \frac{n}{2}(1-\beta)$$

Since number is conserved, we have $n_0 L_0 = nL$ where zero subscript indicated the situation in which the standing wave is stationary. Now $L = \sqrt{1-\beta^2}L_0$ so we finally have

$$n = n_0 \frac{1}{\sqrt{1-\beta^2}} \quad \text{and} \quad n_\rightarrow = \frac{n_0}{2}\sqrt{\frac{1+\beta}{1-\beta}} \quad n_\leftarrow = \frac{n_0}{2}\sqrt{\frac{1-\beta}{1+\beta}}$$

The redshifts satisfy

$$\upsilon_\rightarrow = \frac{1+\beta}{1-\beta}\upsilon_\leftarrow \quad \text{so let there be some frequency } \upsilon_0 \text{ such that}$$

$$\upsilon_\rightarrow = \sqrt{\frac{1+\beta}{1-\beta}}\upsilon_0 \quad \text{and} \quad \upsilon_\leftarrow = \sqrt{\frac{1-\beta}{1+\beta}}\upsilon_0$$

134

The energy density of the forward wave equals $n_\rightarrow h\upsilon_\rightarrow = n_0 h\upsilon_0 \dfrac{1}{2}\dfrac{1+\beta}{1-\beta}$ and

similarly, $n_\leftarrow h\upsilon_\leftarrow = n_0 h\upsilon_0 \dfrac{1}{2}\dfrac{1-\beta}{1+\beta}$ for the backward wave. The amplitudes are

thus proportional to

$$A_\rightarrow = \frac{1}{2}\frac{1+\beta}{1-\beta} \quad \text{and} \quad A_\leftarrow = \frac{1}{2}\frac{1-\beta}{1+\beta}$$

So defining $F_\rightarrow = \exp[i(k_\rightarrow x - \omega_\rightarrow t)]$ etc., we can write

$$\Psi = A_\rightarrow F_\rightarrow + A_\leftarrow F_\leftarrow = SF_\rightarrow + DF_\rightarrow + SF_\leftarrow - DF_\leftarrow = S(F_\rightarrow + F_\leftarrow) + D(F_\rightarrow - F_\leftarrow)$$

where

$$S = (A_\rightarrow + A_\leftarrow)/2 = \frac{1}{2}\frac{1+\beta^2}{1-\beta^2} \qquad D = (A_\rightarrow - A_\leftarrow)/2 = \frac{1}{2}\frac{2\beta}{1-\beta^2}$$

Then from what has been derived in the main text (page 30, ff.), we have inserting, $\gamma = 1/\sqrt{1-\beta^2}$,

$$\Psi = 2\exp[i(\beta\gamma k_0 x - \gamma\omega_0 t)]\{S\cos[\gamma k_0 x - \beta\gamma\omega_0 t] + i\,D\sin[\gamma k_0 x - \beta\gamma\omega_0 t]\}$$

Let $\theta = \beta\gamma k_0 x - \gamma\omega_0 t$, $\phi = \gamma k_0 x - \beta\gamma\omega_0 t$, $\cosh(u) = \dfrac{1+\beta^2}{1-\beta^2}$, $\sinh(u) = \dfrac{2\beta}{1-\beta^2}$

We can re-write Ψ in the form

$$\exp i\theta\,[\cosh u \cos\phi + i\sinh u \sin\phi] = \exp i\theta\,\cosh(u + i\phi) =$$
$$= \exp i\theta\,[\cos^2\phi + \sinh^2 u]^{1/2}\,\exp i[\tan^{-1}(\tanh u \tan\phi)]$$

Finally then, reverting to the original variables,

$$\Psi = \exp i(\beta\gamma k_0 - \gamma\omega_0 t)\otimes$$

$$\otimes\left\{\cos^2[\gamma k_0 x - \beta\gamma\omega_0 t] + \left(\frac{2\beta}{1-\beta^2}\right)^2\right\}^{1/2}\exp i\tan^{-1}\left(\frac{2\beta}{1+\beta^2}\tan(\gamma k_0 x - \beta\gamma\omega_0 t)\right)$$

It is clear that the arguments of the wave-like factor above and that of the envelope factor below are identical to those in the simplified version in the main text, so the arguments made there apply without change to the more exact formulation above.

Part V. Personal notes regarding efforts to publish the variable rest mass interpretation of gravity
(Written February 8, 2006)

May 10, 1977: Satori
This is a brief history of my efforts to gain the attention of the physics community regarding a simple but important insight which came to me sometime around May 10, 1977. On that date I wrote, "I guess that the only conclusion to be drawn is that the mass of bodies is not constant in a gravitational field."

This followed, I reasoned, from two facts: First, since the gravitational red shift must apply to all phenomena, I concluded that the energy (as measured by a distant observer) of a photon pair resulting from the annihilation of a particle/antiparticle pair must depend upon the gravitational potential at the location at which the annihilation occurs: Second, that the energy of a photon does not change as it moves up or down in a gravitational field. This key idea I inferred from the argument that Einstein presented in his brilliant paper of 1911. In this paper Einstein elaborated on his Principle of Equivalence (first enunciated in 1907) and applied it in a remarkable *Gedankenexperiment* to deduce several important results, including the gravitational red shift.

The present writing is not intended to present the theory itself: only the history of my attempts to gain the attention of the Physics community regarding what I consider to be an important insight into the nature of General Relativity. An early presentation of the theory entitled "Toward a Deeper understanding of General Relativity" is still available on the internet.

Roger Babson's Gravity Research Foundation
The first paper that I wrote was entitled "A Causal Analysis of Gravitational Time Dilation and Related Matters." It was submitted in March of 1978 to the Gravity Research Foundation as an entry in its annual essay contest. This institution was founded by Roger Babson, an eccentric who may have had anti-gravity devices in mind when he ponied up the money, but somehow the institution fell into the hands of legitimate scholars, proving that anything, however improbable, occasionally happens.

In this short paper, arguments were presented identifying the change of rest mass as the cause of gravitational time dilation (which is more appropriately called gravitational clock slowing). The most interesting part of the paper dealt with the related matters, namely the fact that the reduction of rest mass causes a dilation of atomic dimensions, and hence an increase in the length of measuring rods. This dilation of measuring rods implies that a profound difference exists between the usual (***proper***) geometry, as defined by strictly local measurements, and the new ***telemetric*** geometry defined in terms of echo ranging based on time signals derived from the 'clock at infinity.' In particular, the geometry of a black hole is changed radically. It was argued that the event horizon of the black hole is in fact infinitely distant in a completely new and foreign universe, the geometry of which is asymptotically one of constant negative curvature. This 'innerspace' universe was shown to be connected to our own familiar universe through a structure resembling what has been called a 'wormhole,' the narrows of which is a spherical surface, which, in spite of its topology, is quite flat! At the very end of the paper, I went on to speculate that it is more reasonable to say that the universe is not expanding, but rather that we, along with our measuring rods, are contracting. It will not surprise you to learn that I was not among the winners, nor among those mentioned honorably.

In April of the same year, 1978, I presented essentially the same paper in a talk to the Cleveland Philosophical Club. For reasons that escape me, I gave it the outrageous title, "The Schwarzschild Solution: After Sixty-Two Years, Satori!" Maybe I had just encountered this wonderful Japanese word denoting a sudden and profound under-standing. This talk established my reputation for delivering incomprehensible talks with great enthusiasm.

Analyzing the Kerr solution

During 1979 and 1980 I made an intensive study of the Kerr solution, which describes the fields of rotating bodies, including those of rotating black holes. I finally succeeded in separating the dynamical effects of 'frame dragging' from effects that are purely gravitational. To do this it was necessary to discover, for each position, a velocity that will maximize the rate at which a clock, moving with that velocity, will run. The residual clock slowing was then assumed to result from the reduction of rest mass at the location under consideration. This in turn determined the factor by which measuring rods were dilated, thus establishing the relation between proper and telemetric geometry. As a proof of the correctness of this approach, I calculated the location of the famous Kerr photon orbits by assuming that these will lie on circles whose *telemetric* curvature matches the curvature induced in the ray paths by the shear in the 'frame dragging' velocity. The results were exactly those calculated directly by the usual methods. I was quite satisfied with this result, but nevertheless I let time slip by, and before I knew it, four years had passed.

A brief exchange with Chandrasekhar

It was in the spring of 1983, I think, that the great Subrahmanyan Chandrasekhar gave the Michelson Lecture at CWRU.[65] He had just recently finished his monumental tome, *The Mathematical Theory of Black Holes*, so naturally his topic was black holes. In fact, one may say that Chandrasekhar, with his prediction of the inescapable collapse of large stars, may appropriately be credited as the discoverer of the modern theory of black holes: so it truly was his topic. It was a marvelous talk, marred only by a rather rude question put to the great man at the very end, when everyone else had gone. I asked him if there was any way that one could understand, in a gut-satisfying way, why the Kerr photon orbits were located as they were. He looked at me with an  expression of slight annoyance, and then replied, "Young man, you have asked me a technical question, so I will give you a technical answer: That's the way it turns out!" His perfect irony was a pleasure for both of us, but I was doubly delighted, smug in my knowing something that perhaps the great man did not.

Marek Abramowicz

The next significant event occurred in 1993 when I read in the March issue of Scientific American an article written by Marek Abramowicz entitled "Black Holes and the Centrifugal Force Paradox."

Way back in 1974 Abramowicz and Jean-Pierre Lasota published their discovery that centrifugal force behaves in a most counter-intuitive manner in the very near vicinity of a black hole: at a certain distance it vanishes, and inside that distance it actually reverses and acts inwardly! The 1993 SciAm article elaborated on this phenomenon and made reference to a theorem due to Brandon Carter, Abramowicz, and Lasota.

---

[65] Note added 2010. Actually, the lecture was occasioned by CWRU's award of the Michelson-Morley Prize to Chandrasekhar.

This theorem states that the paths followed by light rays are the straightest possible paths for particles of non-zero mass, in the sense that the force required to keep such a body on such a path is independent of the velocity of the body. That is to say, inertial 'forces' such as centrifugal force and Coriolis force, whose magnitudes are velocity dependent, vanish on such paths.

Optical Geometry

By analogy to Newtonian mechanics, in which such inertial forces vanish for bodies moving in straight lines, Abramowicz and his associates defined a new geometry for general relativity by identifying light rays as the geodesics (the nearest thing to a straight line in non-Euclidean geometry) of the new geometry. This new geometry they named, appropriately, 'Optical Geometry.' For static situations, it is identical to my Telemetric Geometry.

I was simultaneously elated and slightly disappointed to learn these things. On the one hand, I instantly recognized that my theory implied a geometry in which the reversal of centrifugal force was perfectly obvious. So the behavior of centrifugal force was a powerful confirmation of the correctness of my theory. On the other hand, I was a bit chagrined that I had not myself long ago inferred this now obvious fact.

Submissions to *American Journal of Physics* and *Foundations of Physics.*

This article inspired me to attempt to get my ideas published, and by late 1995 I had completed a very long paper (87 pages in fact). In January 1996 I submitted this monster to the *American Journal of Physics*, a middleweight publication directed toward physics teachers and the interested public. It was of course rejected. I guessed that my inclusion of the 'we are shrinking' cosmological speculation hadn't helped, so I eliminated that part when, in June, I submitted the paper to the journal, *Foundations of Physics*. No surprise, rejected again, I think without any real review. Eighty-seven pages! What was I thinking?

In October 1996 I importuned Glenn Starkman of Case to look at this paper. He was polite but didn't accept the idea that photons move with constant frequency in a stationary field. That is to say, he couldn't accept that any alternative system of measurement (such as the telemetric system) was acceptable, nor, for that matter, needed.

The CWRU Cosmology-Topology Workshop

In September of 1997 Lawrence Krauss and Glenn Starkman of Case Western Reserve University staged a remarkable workshop dealing with the exotic possibility that the universe may be a closed space of negative curvature. All manner of great physicists and geometers were present, but nevertheless, I had the temerity to crash the party. It would be great fun and quite a challenge to try to describe what went on, but memory fades, and in truth I understood only a small part of the presented material. (See the April 1999 SciAm article by Jean-Pierre Luminet, Glenn Starkman and Jeffrey Weeks.)

The kindness of Gary Gibbons

Quite by chance, I happened to be seated next to one of today's great men of general relativity, Gary Gibbons of Cambridge. At the time, I knew nothing about him or his stature (he's listed directly below Stephen Hawking in the DAMTP directory). At a coffee break we chatted, and something came up that bore upon my theory.

I mentioned my work, and Gibbons expressed an interest. He took the whole 87 page thing back to Cambridge, and in a couple of weeks, I received an e-mail that was most encouraging. He said in part, "as an interpretation of standard general relativity I have no problem with it and I certainly feel that using a telemetric interpretation provides many insights."

Importuning Martin Gardner

Thus encouraged, I renewed my efforts, deciding to try for publication in the popular press. Targeting SciAm, I tried to imitate their style, and boiled the whole thing down to eight pages, omitting only the long, detailed derivation of the Kerr photon orbits. I had a previous connection ('Dragon Curves') with Martin Gardner, former well-loved editor of the 'Mathematical Games' feature of SciAm, so I sent him the new paper, hoping that he would use his influence to get it into the magazine. He claimed to admire the proposed article, but he felt that it was far too technical for SciAm. He made no offer to influence the editors.

I gave a copy of this short paper to Lawrence Krauss of Case. Polite silence.

On January 1, 2000, I posted on the Los Alamos website (then, xxx.lanl.gov: now, arXiv.org) this short paper titled 'Toward a Deeper Understanding of General Relativity'. It may be found there today.

Abramowicz invites me to a workshop on 'Optical Geometry' in Trieste

Finally, I reasoned that if anyone in the legitimate physics community would be likely to accept my ideas, it would be Marek Abramowicz, since we already agreed on the geometry of black holes. In May 2000 I sent an e-mail to Abramowicz, asking him to look at my paper on the Los Alamos website. He replied immediately, saying that we were in full agreement regarding the geometry of the static black hole, but rejecting my idea of variable rest mass. However, he went on to invite me to participate in a workshop on Black Holes and Optical Geometry that he and Antonio Lanza were organizing to be held at the International Center for Theoretical Physics in Trieste, Italy, 15 August through 15 September 2000. I accepted enthusiastically.

The ICTP is situated just north of the city of Trieste on a steep hillside adjacent to a spacious, lovely parkland that surrounds the elegant Miramare Castle, which is perched on a high bluff over the Adriatic. (For a virtual tour use the following link: <http://www.ictp.trieste.it/pages/info/virtualGRN.html>

Just north of the castle lies the townlet of Grignano with its little harbor, string of restaurants, and the Guesthouse Adriatico, my home for two weeks. It was delightful. The Adriatico is operated by SISSA, the Scuola Internazionale Superiore di Studi Advanzati, a sister institution of ICTP, sharing the same hillside campus.

I was surprised to discover that there were very few people involved in the initial sessions of the workshop. A few more appeared later. Here's a list of those who were there during my stay.

Marek Abramowicz, Chalmers and Göteborg Universities, Sweden
Pawel Nurowski, Warszawa University
Hans Westman and Rickard Jonssen, students of Abramowicz
C.V. Vishveshwara, Indian Institute of Astrophysics, Bangalore
G.F. Torres del Castello, University of Puebla, Mexico
Stefano Liberati, SISSA
Vladimir Karas, University Karlova, Prague

At our first meeting, Abramowicz made a point of repeating his stand regarding my paper: He conceded that I had the geometry correct, but he didn't see any need to consider how fundamental constants, such as the mass of the electron, might be affected by gravitational fields. The first sessions were mostly devoted to laying out a schedule for the workshop. After two days, Abramowicz flew back to Göteborg for administrative meetings at his university.

While he was away, we listened to Pawel Nurowski discuss techniques for the separation of variables in the case of nonlinear equations such as Einstein's field equations. I also presented two papers: one on the basis of the variable rest mass interpretation, and a second on the very speculative cosmological implications of the theory. I was saving my best argument for my theory, the analysis of the Kerr black hole, for the return of our leader. (This analysis was later posted to the arXiv website.)

To my great surprise, as soon as he returned, Abramowicz told the group that he had been thinking about my variable rest mass approach, and was now inclined to believe that it was correct. It was then a pleasure to present my treatment of the Kerr solution (the rotating black hole), for I felt certain that this would clinch the case for my interpretation, since for the Kerr, my approach succeeds, where Optical Geometry fails. Let me briefly describe why OG fails.

For a rotating black hole, it is now common for relativists to say that space itself is swirling around the hole, dragged, as it were, by the rotation of the hole. This is not the best way to describe the situation, but it will do for our purposes. The important thing is that the speed of rotation of space (or, I would say, the aether) increases as one approaches the hole. In such a situation, a light ray moving tangentially will be turned by the shear in the flow. If the ray is directed opposite the flow, the wavefront of the ray will be retarded more on the inner side than it is on the outer side, and the ray will curve toward the hole. On the other hand, if the ray is directed in the same direction as the flow, the inner side of the wavefront will be speeded up more than the outer side, and the ray will curve away from the hole. This makes it clear that in such a situation, the light ray path from A to B is not the same as that from B to A. Thus it is not possible to associate geodesics with the path followed by light rays, for the latter are direction dependent. And so the optical geometry approach of Abramowicz, Carter and Lasota, fails in this case.

A tepid response to my presentation
My presentation was accepted with an absence of enthusiasm. Everyone appeared to understand what I was saying, and no one claimed that errors had been made. They all acted as though it was a matter of choice, and they didn't choose to embrace my interpretation.

The remaining sessions of the workshop were very interesting but not related to the variable rest mass interpretation of general relativity. A major topic was the characterization of the various inertial forces (the ones that vanish on light ray paths in static fields). Hans Westman and Abramowicz had an animated disagreement about the right way to do this. And, of course, the workshop continued for two weeks after my departure.

Abramowicz: "I am now more at your side about the variable mass."
On October 25th 2000, I sent an e-mail to Marek Abramowicz thanking him for his kindness in inviting me to participate in the workshop, and requesting that he give me his assessment of the papers that I had presented. I made it very clear that in the case of the Kerr, I considered that my variable rest mass approach succeeded where the optical geometry approach failed. Here's his response, received the same day:

Dear Jack,

It was great to have you at the workshop. I am now more at your side about the variable mass. I am back in Sweden, but next week I go to Scotland and Oxford, and just before Christmas I will be at the Texas Symposium, giving one of the major talks, this time not on OG, but accretion disks.

Best regards,
Marek

A small, but very sweet success.

In November of 2000, as a result of ongoing kinematic studies by Abramowicz and his graduate student, Hans Westman, the tide seemed to turn, and the current of opinion appeared to switch in favor of my approach.

Sonego and Abramowicz

Sometime later I discovered that another colleague of Abramowicz, Sebastiano Sonego of the University of Udine in Italy, had been for some time advocating the variable rest mass idea. His approach was different from mine, in that he proposed to introduce, ad hoc, a gravitational potential that affects only particles of non-zero rest mass, leaving photons and other massless particles unaffected. In my analysis, the introduction of such a potential is not required, since the standard theory already has these features. Late in 2000, Sonego and Abramowicz planned a paper on Optical Geometry, tentatively titled, "The hidden geometry interpretation of general relativity," but other more pressing matters apparently interfered. About the same time, Abramowicz and Sonego announced the publication of a new book titled, *Black Hole Physics in the Optical Space*, but publication has also been delayed. [Note added 1/28/2010: after several postponements, publication has been cancelled.]

Submission to *Phys Rev D*: Exchanges with Erick Weinberg and final rejection

Encouraged by my interaction with Abramowicz and Sonego, I determined to try again to get the paper published in a refereed journal. I first tried *Phys Rev D*.

The editor, Erick Weinberg, dismissed the paper without review by referees, indicating, in what appeared to be a form letter, that "reinterpretations ... which lie outside the mainstream ... and are not at the level of discourse appropriate to the current state of research in the field..." could not be accepted. He was, nevertheless, kind enough to respond to questions that I posed:

(1) Do you think that it is unnecessary to try to discover a *cause* for the gravitational red shift? If so, why do you believe that this phenomenon is uniquely immune to investigation?

(2) It is well known that the time required (as measured by any stationary observer) for an object to reach the event horizon of a black hole is infinite. Doesn't it then follow that nothing has ever penetrated, or ever will penetrate the horizon?

(3) What is your explanation for the reversal of centrifugal force inside the photon orbit at 1.5 times the Schwarzschild radius?

Professor Weinberg responded as follows:

With regard to questions one and three, I would say that the explanations are provided by the equations of general relativity. The red shift and the properties of photon orbits at small radii [Weinberg seemingly misunderstood question three] are mathematical consequences of these equations. One can look for heuristic arguments as to why these are the solutions, but the equations themselves are the explanations. The answer to your second question lies in recognizing that English grammar assumes a Galilean geometry. In a more complex geometry, such as a Schwarzschild geometry, phrases such as "nothing has ever penetrated" need more explanation. A correct statement is that (1) the past of any observer outside the horizon does not include the crossing of the horizon by any object and (2) the past of any observer inside the horizon can include the crossing of the horizon by objects.

In his answers two attitudes are strikingly evident. First, in seeking understanding, one need look no further than the mathematics – "*the equations themselves are the explanations.*" Second, we see the adherence to the dogma that all observers must be accepted as being equally credible, even if contradiction must be accepted.

I have to give him credit for his courtesy in replying to these somewhat impertinent questions, but his neo-positivism is baffling to me. How can a scientist not be interested in the cause of phenomena as striking and counter-intuitive as the slowing of clocks, and the astounding reversal of the direction of centrifugal force? Guessing the motivations of others is not just liable to be wrong, but it is also presumptuous and usually self-serving. But still, one wonders why brilliant people can embrace a view of reality so foreign to the natural inclination of man, to **understand**. I would guess that it has to do with the struggle as an undergraduate to understand relativity (as it is usually presented) and quantum mechanics. Here I quote from the introduction to Chapter 4, (titled "Relativity") of Paul Richards' book, *Manual of Mathematical Physics:*

> This fact [that all observers, regardless of their relative motion, will insist that a single pulse of light remains centered on a particular point in their particular system] is completely foreign to intuition and cannot be reconciled with one's natural experience. The only course is to abandon intuition and investigate what this strange but inescapable *fact* implies. [Emphasis Richards']

Regarding quantum mechanics, there's Feynman's famous advice,

> I think it is safe to say that no one understands quantum mechanics. Do not keep saying to yourself, if you can possibly avoid it, 'But how can it possibly be like that?' because you will go down the drain into a blind alley from which nobody has yet escaped. Nobody knows how it can be like that.

So it seems to me that perhaps those who are able to continue in physics in spite of a lack of any natural understanding tend to protect themselves by denying the very possibility of such understanding. Maybe there is also some satisfaction in scoffing at the naïve 'pictures' that amateurs propose.

Submission to *Foundations of Physics*, and a curt dismissal from the editor, Alwyn van der Merwe

Subsequent attempts to publish in a refereed journal proved that Weinberg's attitude was pervasive among editors, and, one must assume, among physicists generally. I next tried *Foundations of Physics*, and received a curt dismissal from the editor, Alwyn van der Merwe, who, in refusing to answer the same sort of questions, wrote

> I trust that you will understand that I, laboring as I do 12 to 15 hours a day, seven days a week, while trying to meet unforgiving publication deadlines for two international journals and a 120-volume bookseries, in addition to pursuing my own research and writing (which presently includes two books) and fulfilling my teaching duties--all without the help of a single typist or secretary--do not have the luxury of engaging in discussions with authors about the substance of their papers, which would be a suicidal use of my time.

Last try: *General Relativity and Gravitation*

Most recently, I submitted the paper to *General Relativity and Gravitation*. The paper was swiftly rejected ("the paper contains nothing really new"), but when I pleaded my case, it was sent out for 'adjudication.' After a couple of months, the adjudicator, complaining that the arguments presented "do not make sense to me," recommended against publication.

A final review by Robert M Wald

It is the policy of *GRG* that a final review, if requested, will be done by the chairman of the Society, at that time, Robert M Wald. Professor Wald was very prompt and thorough, but again, he held to the same epistemological attitude that Weinberg had expressed. Thus, in response to the same sort of questions, he said:

> I agree with your quote of Weinberg to the extent that I believe that a well posed theory, such as general relativity, does not require any additional "explanations" of its predictions. However, I also agree with what I would take to be your view that "explanations" can be useful both for gaining intuition into the nature of the predictions made by the theory and, most importantly, for identifying structures that might appear in a generalization of the theory that may eventually supercede the original theory. It is certainly the case that what constitutes a useful "explanation" of a prediction of a theory is largely a matter of taste and judgment, since, in particular, what may seem like a "counterintuitive" prediction to some people may not appear so to others who have more experience with the theory. It is my general view that any "explanation" that involves the introduction of additional structure that is well defined only in a restricted class of examples is not likely to be useful either for gaining intuition or for pointing the way towards a generalization of the theory.

It is true that the proposed scheme of measurement can only be implemented in situations for which a world time can be defined, but this does not mean that the idea that rest mass is reduced in a gravitational field is not generally true. And the situations in which the scheme *can* be carried out include the two most important spacetimes: the rotating black hole, and the Friedman spacetimes of cosmology.

In a further exchange, I wrote

> I would conclude from your general statement that you do not see any compelling reason to pursue an investigation of the possible cause of gravitational clock slowing. The more usual name for this phenomenon is 'gravitational time dilation,' and one often sees the statement that 'time itself passes more slowly' as one descends in a gravitational field. I am wondering whether you subscribe to this 'relativity of time,' which, if accepted, would render any inquiry as to the 'cause' (quotes now appropriate) absurd.

146

Professor Wald answered

> To me, the question is analogous to asking why the distance between longitude lines is smaller near the north pole than the equator. I am satisfied with "the surface of the Earth is curved" as providing the explanation, and I do not see the need to seek any additional "cause." Others might feel differently.

A shot fired in anger

This anti-analytic attitude seems to pervade the thinking of physicists these days. I was unable to let go quietly, as perhaps I should have done. On April 8th 2003, I fired off the following e-mail

> Dear Professor Wald,
>
> Since you were kind enough to enlighten me on certain matters in GR, I thought it appropriate that I should return the favor.
> In section 6.3 of your otherwise superb text, *General Relativity,* you make the following comment regarding the gravitational red shift: "Physically, this makes sense because according to quantum theory the energy of a photon is proportional to its frequency, $E = h\nu$, and we would expect the photon energy to be degraded as it 'climbs out of the gravitational potential well'." But of course, energy is conserved in a static field such as the Schwarzschild field. (cf. Schwinger, *Einstein's Legacy*. p.142)
>
> Again in the same section, you remark, "...there will be a nonzero deflection of the light ray, which we may interpret physically as being due to the gravitational attraction of the Schwarzschild geometry." That light rays suffer no gravitational attraction is evident from the fact that photon orbits occur precisely at that location where centrifugal force vanishes: at 3/2 the Schwarzschild radius.
> It is my opinion that the insights provided by the VRM interpretation would preclude such misunderstandings. Let me know if I may be of further assistance.
>
> Sincerely,   Jack Heighway

Professor Wald did not respond. I am now convinced that further attempts to publish in a refereed journal would be futile.

The Positivists are not consistently anti-heuristic

The reader will have noticed that despite the assumption of a purist positivist attitude in rejecting my paper, Professor Wald, and indeed nearly all of the many physicists writing texts, do, in fact, attempt 'physical' explanations for the red shift and other phenomena of GR. Perhaps such heuristics are judged to be acceptable in a textbook, but not in a journal article. Or perhaps the objections to such explanations merely serve as a cover for unstated objections – the author's non-traditional style, his lack of credentials, or perhaps the out-of-the-mainstream conclusions of the paper itself.

Also, it is easily imagined that these editors and referees might feel certain that nothing so simple and obvious, if it were in fact true, could have possibly remained undiscovered for such a long time.

Summary of the variable rest mass interpretation of gravity

It seems appropriate here to review the situation as it exists today and to summarize the main features of the new Variable Rest Mass (VRM) interpretation of general relativity interpretation offered in the subject paper.

Today, the most influential physicists seem unwilling to concede that energy levels are reduced in a gravitational field, let alone accept the idea that rest masses are similarly affected. The reason for this, one can only guess, is that modern day physicists are committed to a neo-positivist view, according to which one must not formulate theories that go beyond immediate observation. At the same time, the notion that all inertial observers are to be accepted as equals seems to have been carried over from the special theory of relativity, and applied to (non-inertial) observers in a gravitational field. But as we have seen, observers at different levels in a gravitational field each have different 'standards' of measurement. Somehow, evidently in the spirit of philosophical relativism, modern day physicists accept this chaos. They are even willing to accept the absurd idea that 'time itself evolves differently' for each observer. It would appear that specialists in general relativity believe that nothing can be done to improve upon this 'tower of Babel' scheme for measurement. But they are wrong.

One *can* introduce a new system of measurement that is not influenced by gravity. It is the constancy of the frequency of freely moving electromagnetic waves that makes possible the implementation of this new system of space-time measurement. One simply measures time using the signals from a single remote clock. Distance measurements are then made (by local or remotely located observers) using electromagnetic echo ranging (radar) techniques, calculated using the time as measured, not by a local clock, but by the same remote clock. Since the new system of measurement employs a remote clock and makes use of a remote sensing technique for distance measurements, I have suggested that it be called the telemetric system.

It must be admitted that this telemetric scheme cannot be implemented in a completely arbitrary spacetime. It is required that the spacetime be time-independent, and asymptotically static at infinity, (or, as in the cosmological situation, spatially homogeneous and isotropic). But this restriction includes a most important solution, the Kerr spacetime, which is believed to describe rotating black holes. It seems obvious that if rest mass reduction occurs in one important class of spacetimes, it is reasonable to believe that it is a universal phenomenon, even though it does not appear to be feasible to define a gravity-independent system of measurement in the case of an arbitrary spacetime.

What are the implications of this Variable Rest Mass interpretation of general relativity? First of all, the VRM interpretation clarifies the connection between general relativity and Newtonian theory, providing a satisfying answer to the question regarding the nature of gravitational potential energy. It explains why light rays do not follow the path which proper measurements purport to be the shortest path. It explains the Shapiro effect (the delay of radar echoes from a body near superior conjunction with the Sun) without resorting to the ad-hoc assumption that the speed of light is reduced in a gravitational field. Perhaps most impressively, the mind-boggling reversal of the direction of centrifugal force (the Abramowicz effect) very near a black hole (i.e., for $r < 1.5 \, r_S$) becomes understandable.

Regarding black holes: As assessed in the telemetric system, a collapsing star does not suffer an infinite compression, producing a singularity hidden behind an event horizon; when its proper circumference becomes less than $(9\pi/4) \, r_S$, it produces a sort of rupture in space, forming a connection to a distinct infinite realm, which may be called 'innerspace,' into which it falls. The worrisome 'event horizon' at r_S is actually infinitely remote in 'innerspace,' and the region within (beyond) it has exactly the same insignificance as the region 'beyond infinity' has in our familiar universe.

In addition, the imagined problem of the loss of information associated with the disappearance of matter falling through the event horizon at r_S is solved, since it becomes clear that nothing has ever fallen, nor will ever fall, through the event horizon, since that hypothetical 'surface' is actually infinitely distant. For the same reason, the fearful singularity inside the event horizon, at $r = 0$, is no longer of concern because of its non-existence in our universe.

More speculatively, the VRM interpretation has also been extended to the cosmological problem, leading to the seemingly silly conclusion that the universe is not expanding, but rather that all of our devices for measuring distance (measuring rods, radar based upon local time scales, etc.), and indeed we ourselves, are contracting! Here also, the new interpretation has several advantages over the standard interpretation: it re-connects with Newtonian concepts, conserving momentum in terms of the usual definition, and it provides an explanation for the cosmic red shift that does not involve the rather mysterious concept of the expansion of space itself. Perhaps the best features of the interpretation are those that are not present, namely, the infinite density and singularity of the big bang, and the faster-than-light expansion of space itself, which, in the standard interpretation, is an unavoidable feature of inflationary theories.

Perhaps Ptolemy, in the improbable hereafter, may have objected to the new theory of Copernicus saying, 'But it's merely an interpretation!' It was merely that, but nevertheless it has had a profound effect on our thinking.

Jack Heighway   February 8, 2006

Afterword

I would first like to emphasize the importance of certain transformative ideas that have been presented in this book.

Regarding special relativity, it has been demonstrated that all seemingly counterintuitive phenomena result from a failure to recognize the implications of Einstein's prescribed technique for the synchronization of clocks. This scheme is based upon Einstein's unwarranted 'definition' that the time required for light to travel from A to B is the same as that for travel from B to A. A proof of the existence of the aether is presented, in terms of which simple explanations of seeming puzzling phenomena are given.

Regarding the theory of gravity, proof has been presented that variable rest mass is actually inherent in Einstein's theory. Careful consideration of the gravitational red shift (aka the gravitational slowing of clocks) proves that the rest masses are reduced in gravitational fields. This variable rest mass interpretation makes clear the true nature of gravitational potential energy. When an object is raised, the increase in gravitational potential energy is real, and exists as an increase in the rest mass of the object, in accordance with the equation, $E=mc^2$. Thus if work, ΔE, is done raising a body, the rest mass of that body will increase by the amount, $\Delta m = \Delta E/c^2$. Conversely, lowering an object (and absorbing the released potential energy) will result in a decrease in the rest mass of the object. This obvious and intuitive concept replaces the vague and unproven assumption that gravitational potential energy is manifest as a change in the negative energy of the gravitational field, the field energy becoming less negative when an object is raised.

Importantly, this reduction of rest masses also implies that the true dimensions of all material objects are increased by the same factor by which clock periods are increased. This in turn implies that the currently accepted geometry is incorrect, since it is, in effect, based on measurements employing dilated measuring rods.

The same analysis of the gravitational red shift also proves that light, moving freely in a time-independent gravitational field, does so with no change in frequency, contrary to commonly held opinion.

This fact enables a system for the measurement for time and distance that is uninfluenced by gravity. This 'telemetric' system gives the correct description of the geometry of space in a gravitational field. This geometry explains the otherwise incomprehensible Abramowicz effect, the reversal of centrifugal force near a black hole, and puts to rest the crisis posed by the feared loss of information and entropy that arguably would occur as a consequence of material crossing the event horizon. This 'surface' is in fact utterly unreachable, being the infinity of 'innerspace.'

Although the frequency, and hence the energy, of light is unaffected by gravity, light is *indirectly* influenced by gravity in that it exists in the non-Euclidean geometry associated with the gravitational field. Light moves along the geodesics of the true geometry as described by the telemetric system.

That light is unaffected by gravity is a fact of inestimable importance. It means that mass and energy are not in every way equivalent. ***It means that gravity acts only on objects with rest mass!*** The implications of this fact are deep and far-reaching. In particular, it means that the massless gravitons quantizing the gravitational field do not interact with that field, nor with one another, implying that the quantum theory of gravity need not be intractably non-linear.

It is widely accepted that gravity, being characterized in Einstein's 1915 theory as a tensor field, is to be quantized by a corresponding spin-two quantum particle. But when the straightforward procedure for quantizing a field is followed, one gets a mixture of spin-zero and spin-two gravitons.

Now spin-zero particles can only interact with fields whose energy-momentum tensor has a non-vanishing trace. But in the case of the electromagnetic field, the trace of this tensor field *does* vanish. For this reason, physicists, believing incorrectly that electromagnetic fields *do* interact with gravity, have resorted to a rather contrived mathematical artifice to eliminate the spin-zero graviton.[66]

But as we have seen, light does *not* interact with gravity. Thus, it's my guess that the instantaneous, static component of the gravitational field is mediated by a spin-zero graviton, while the dynamical component of the field, gravitational waves, are quantized by spin-two gravitons.

Applied to cosmology, the variable rest mass interpretation gives rise to a radically different interpretation of the classic Robertson Walker solutions to Einstein's gravitational field equations. The function that is conventionally interpreted as a scale determining the increasing distance between galaxies is, under the new interpretation, identified as a function determining how rest masses evolve, increasing with the passage of time. An analysis based upon the conservation of momentum, (which is required by the assumption of the large scale homogeneity of the universe) indicates that rest masses have been, and are, increasing as a simple exponential function of world time, t, (the 'flow' of which is unaffected by changing rest masses), in terms of which the big bang occurred at $t = -\infty$, corresponding to a finite 'proper time' on the order of 10^{10} $years$.

The increase of rest masses with time implies that the red shift has nothing to do with the expansion of space itself (the current understanding), but rather is a simple consequence of the fact that in the past, spectral emission frequencies were reduced because of the reduced rest masses of the era in which the emission occurred. The increasing rest mass of all matter also implies that our measuring rods are contracting, leading to the surprising conclusion that the universe is not actually expanding. The whole character of the 'big bang' is radically different from the currently held model: the singularity is absent and the temperature remains finite as the rest masses of all matter fields going back in time tend to zero.

---

[66] Richard Feynman (1995) *Feynman Lectures on Gravitation* Addison Wesley, p.35-38

Finally, I can imagine that some readers may feel that I have treated Einstein rather roughly. It is not the man, but rather certain parts of his work that I have criticized. No thinking person can be but awestruck by Einstein's contributions: the bold analysis of the photoelectric effect introducing the photon concept; his inspired and 'out of the blue' theory of gravity; his resolute defense of realism in his ingenious presentation of the EPR paradox, which led to Bell's theorem and Alain Aspect's experiment establishing the non-local nature of quantum interactions; and finally, his steadfast and courageous advocacy in support of peace and justice. No man has done more.

Nevertheless, criticism could not be avoided in the treatment of special relativity, since the treatment presented is based on the Lorentzean aetherist point of view.

Regarding the theory of gravity, it seems clear to me that Einstein's greatest blunder was not his introduction of the cosmical constant, but rather his faulty analysis of the implications of the clock slowing effect, which he had predicted on the basis of his brilliant 'elevator' *Gedankenexperiment* of 1911. He was mistaken in his unwarranted assumption that the speed of light had to slow in proportion to the slowing of clocks, in order that the locally measured speed of light remain constant. Having (seemingly) resolved this crisis, he evidently saw no reason to attempt to discover the cause of the slowing of clocks – and all processes – in a gravitational field. This is understandable since, in 1911, there was no theory to systematically deal with such matters. But after Bohr's 1913 quantum theory of the hydrogen atom, the necessary theory was available. (Personally, I cannot understand why neither Einstein – nor anyone else – tried to discover the cause of gravitational clock slowing. This was not just a failing on Einstein's part: it has been an ongoing failure of the entire physics community.) Finally, Einstein did not explicitly state that light moves with constant frequency in a time-independent gravitational field. Had he done so, the confusion regarding the gravitational red shift would not have developed. But more importantly, it would have been evident that gravity does not affect particles of zero rest mass, and that, conversely, such particles do not act as sources of gravitational fields.

The remarks of Erick Weinberg and Robert Wald quoted in Part V provide an insight into the prevailing philosophical attitudes of physicists. Much of the community has become imbued with a weird combination of logical positivism and mysticism. On the one hand, attempts to develop an understandable explanation of puzzling phenomena are belittled as unnecessary, since "the mathematics says it all." At the same time, bizarre ideas such as Everett's 'many-worlds' interpretation of quantum mechanics; or the idea that consciousness causes the collapse of the wave function; or even the possibility of time travel! – are all blithely accepted as reasonable possibilities.

All this philosophical relativism began with Einstein's rejection of a preferred frame of reference, the aether, and with it, the existence of a universal temporal ordering of events, and the concept of unambiguous simultaneity. Admittedly, an aether rest frame cannot be defined in our universe, but the implications of its existence are vital to the realist point of view, which has been, and should remain, the basis of all scientific endeavor.

Postscript. Frank Wilczek's wonderful book, *The Lightness of Being*, came to my attention only after this book was essentially finished. It changed my thinking regarding the character of the aether most profoundly, and forever altered my mental picture of the nature of the universe.

Wilczek's uses the word 'Grid' to describe the multiple 'aethers' (fields of various sorts, actually), which fill all space, everywhere in the entire universe. The physics of the Grid is strikingly analogous to the physics of the solid state. The concept of superconductivity plays a key role in the interactions of the various fields, giving rise to excitations which we have hitherto regarded as 'particles' existing in 'empty space.'

I have begun the second of doubtless many readings of this remarkable book. It is not to be missed.

J H

INDEX